T0348916

The
Art
of
Physics

The Art of Physics

How Science Explains the Chaos, Contradictions, and Unpredictability of Life

ZAHAAN BHARMAL

GREYSTONE BOOKS
Vancouver/Berkeley/London

25 26 27 28 29   5 4 3 2 1

Greystone Books Ltd.
greystonebooks.com

Cataloguing data available from Library and Archives Canada
ISBN 978-1-77840-274-6 (cloth)
ISBN 978-1-77840-275-3 (epub)

Jacket design by Alex Kirby
Text design by IDSUK (Data Connection) Ltd

Printed and bound in Canada on FSC® certified paper at Friesens. The FSC® label means that materials used for the product have been responsibly sourced.

Greystone Books thanks the Canada Council for the Arts, the British Columbia Arts Council, the Province of British Columbia through the Book Publishing Tax Credit, and the Government of Canada for supporting our publishing activities.

Canada

Greystone Books gratefully acknowledges the xʷməθkʷəy̓əm (Musqueam), Sḵwx̱wú7mesh (Squamish), and səlilwətaɬ (Tsleil-Waututh) peoples on whose land our Vancouver head office is located.

To my parents, Zulfi and Rutton Bharmal

CONTENTS

INTRODUCTION

At the World Economic Forum in Davos, the single most valuable commodity is not money, clothes or the car you drive. It's the colour of your badge. Every year, the global elite – political giants, industry titans and a sprinkling of cultural luminaries – descend on this small city nestled in the Swiss Alps to discuss the most pressing issues facing the world. Access to these discussions is strictly controlled and it's the colour of your badge that determines your status for the week. Prime ministers, presidents and CEOs are given coveted white badges, offering unfettered access to virtually everything. Then, in order of descending importance, there are orange, green, purple, blue and red badges. My badge, however, was none of these colours.

It was gold. A rich and opulent shade of gold that, for a brief moment, I thought might be a secret class of badge reserved for extra-special attendees. Unfortunately, I soon discovered that gold-coloured badges are close to the bottom rung, the Davos equivalent of the 'participation medals' my children might receive at their school sports day. Sometimes just called hotel badges, gold badges are given only to support staff – individuals who are allowed to enter hotels but literally not set foot beyond the lobby. Any attempts to go further – to enter one of the many splendid hotel conference rooms where the important discussions took place – would be met with prompt refusal by security. Nevertheless, collecting my gold badge on a snowy Sunday morning in January 2020, I was incredibly excited.

The World Economic Forum Annual Meeting is never a dull gathering. But 2020 was a particularly fascinating time to attend such an important event. Although Covid-19 would not force most of the world into lockdown for another few months, the virus had already been identified and the World Health Organization was starting to worry. That and other pressing topics, such as climate change, economic precarity and democratic instability, were some of the key topics of debate.

I would not be joining any of these discussions. I was there to support a senior executive – a white-badge holder. This meant my role was diligently organising schedules, writing briefing documents and crafting apt remarks. My job was to get my executive to the meeting room, prepared and on time, and watch the door close in front of me as she walked inside. As much as I would have loved to have entered the room, this arrangement did have one small benefit: as an introvert, it gave me precious time to myself. So while my boss was busy in meetings behind closed doors, I could sit quietly in the hotel lobby on my own, browsing through whatever reading material was lying around. And, at that time, this included the World Economic Forum's 'Global Risks Report'.[1]

Every year, the Davos organisers publish this weighty report to help provide context and frame the forum's various discussions. The report systematically assesses the likelihood and potential impact of the greatest threats facing the world. The 2020 issue described a global economy facing an increased risk of stagnation, and climate change striking harder and more rapidly than expected – all while citizens worldwide protested political and economic conditions and voiced concerns about spiralling inequality. Suffice to say, it was not a fun read. Faced with such epic, intractable, existential challenges, none of which have obvious solutions, I felt quite overwhelmed. I would have liked to have been able to contribute and help in some way but, compared to the brilliant minds convening in Davos, I knew I had nothing special to offer. Well, almost nothing.

---

[1] The World Economic Forum 'Global Risks Report 2024'. [The latest edition at time of writing.] https://www.weforum.org/publications/global-risks-report-2024/.

As I read the impressive biographies of the various white-badge holders attending Davos, I noticed something interesting. As one might expect, the vast majority had advanced qualifications in fields like economics, finance or the political and social sciences. There were experts in medicine, earth sciences and agriculture. There were artists and creators, historians and writers, even philosophers and theologians. But I noticed that relatively few had a background in one particular subject. A subject that is arguably the oldest and most fundamental of all academic disciplines. A subject that has helped reveal the very nature of the universe for over two millennia. A subject that I fell in love with as a boy and went on to study at university. A subject whose lessons and ideas have, in the years since graduating, helped me immeasurably. That subject was physics.

What does physics have to do with issues discussed at Davos? Not much, you might think. Davos is about the messy world of people and politics. Physics is about the methodical world of science and reason. Surely the two could not be further apart. Yet in recent years, many people far more influential than me credit thinking like a physicist with their success in fields far removed from physics. The late Charlie Munger, for instance, attributed physics and its fundamental approach to problem-solving for much of his and his billionaire partner Warren Buffett's extraordinary wealth. In his blog, Dominic Cummings, the divisive mastermind behind the Brexit Leave campaign in the UK, credits his political victory to hiring physicists, as opposed to people from more traditional fields like politics, business or economics. And unsurprisingly, Elon Musk has spoken on numerous occasions about how physics has influenced his many endeavours, from tackling the challenge of sustainable transport to seeking to protect the future of our species by building rockets to Mars. For Musk, physics, more than any other discipline, is predicated on first-principles reasoning and being able to 'boil things down to their fundamental truths'.

And so, sitting in that Davos hotel lobby, I asked myself the questions that ultimately provided the inspiration for this book. Could ideas from physics somehow provide a different way of thinking about global risks? Could physics offer any practical solutions? Ultimately, could physics save the world?

* * *

I haven't always liked physics. In fact, when I was 14 years old and beginning middle school, it's fair to say that I hated it. And it felt like physics hated me too. There were some subjects at school that I found hard and some I found boring. Physics was both. My grades reflected this, and at the end of the first school term my well-meaning but disappointed physics teacher told my parents, with no sense of irony, that my progress 'lacked momentum'. Momentum in physics is defined as mass multiplied by velocity. I quite literally wasn't going anywhere.

The truth is, like many young teenagers, I had other things on my mind. I was an anxious, awkward first-generation Indian, growing up in suburban north London. Attending a prestigious but highly competitive private school, I had few friends and found navigating the complexities of social relationships phenomenally difficult. I was more interested in understanding the mysteries of popularity, not physics. The world felt chaotic and miserable.

But on my fifteenth birthday, things began to change. As part of their ongoing campaign to encourage me to read more books, my parents bought me Douglas Adams's comic science fiction masterpiece, *The Hitchhiker's Guide to the Galaxy*.[2] I immediately connected with Arthur Dent, the hapless protagonist buffeted by forces beyond his control and outside his comprehension, yet still making something of his life. But more than that, I was drawn to the idea of a single 'answer to the ultimate question of life, the universe and everything'. For me, the idea of such

---

[2] *The Hitchhiker's Guide to the Galaxy*, Douglas Adams, Pan Books, 1979.

an answer was not only entertaining, it was fantastically comforting.

I realised, of course, that the concept was just a conceit. Adams was making a joke of it, skewering the search for a single final answer by making it ridiculous. Famously, it turns out that the answer to life, the universe and everything is the number 42. But as a lonely teenage boy, I became enchanted by the possibility of a single framework that could one day help make sense of the world around me. I wanted to find the real version of this framework, and physics – the most fundamental of the sciences, probing the very nature of matter – felt like the path to get me there.

Like all young students, one of the first physicists I learned about was Sir Isaac Newton. In 1665, the Black Death had killed a quarter of London's population. As this devastating outbreak of the bubonic plague spread, Cambridge University closed its doors and a 22-year-old Newton, then a fellow at Trinity College, was forced into the 17th-century equivalent of lockdown. Returning to his childhood home in Woolsthorpe-by-Colsterworth, Lincolnshire, and free from distractions, he embarked on a period of extraordinary discovery. Legend has it that under the shade of an apple tree, a falling fruit sparked a realisation. Newton hypothesised that the force that pulls an apple to Earth must be the same force governing the motion of the Moon in its orbit. This seemingly simple observation blossomed into Newton's three laws of motion.

The first law, or the Law of Inertia, states that an object at rest will stay at rest, and an object in motion will remain in motion with constant velocity unless acted upon by an external force. This law explains why a rocket on the launch pad doesn't go anywhere until its engines fire, and why, once it is on its way to the Moon, it turns its engines off again and coasts on momentum until it needs to slow down at the other end. Inspired by this insight, Newton delved deeper.

His second law is commonly expressed as an equation: $F = m \times a$. Force equals mass multiplied by acceleration. Or put another way, mass equals Force divided by acceleration. This means that the heavier the rocket, the more powerful the engines need to be to get it moving to a velocity that will escape Earth's orbit.

Newton also observed that for every force there was an equal and opposite reaction. This principle became his third law, the Law of Action and Reaction. This is the law that makes rockets work in the first place – the thrust of the exhaust in one direction causes the rocket to move the other way.

These three laws also paved the way for Newton to develop his universal theory of gravitation, which states that every object in the universe attracts every other object with a force that is proportional to the product of their masses and inversely proportional to the square of the distance between them. In other words, the bigger an object's mass, the bigger its gravitational pull (which is why the Earth has stronger gravity than the smaller Moon), and the further away two objects get, the weaker their attraction (which is why the Sun's gravitational pull felt by the Earth is stronger than that felt by more distant planets in the solar system). This elegant expression laid the groundwork for our understanding of the cosmos for centuries to come. It represented a universe that was predictable, ordered and certain. And it also inspired the work of countless other scientists.

One of those scientists was Pierre-Simon Laplace. In the 19th century, Laplace made crucial contributions to understanding planetary motion by applying Newton's theory of gravitation to the entire solar system. But Laplace's real ambition extended well beyond physics. He was driven by a deep desire to predict not just the motion of planets but to formulate an all-encompassing theory of the universe, a theory of absolutely everything. Laplace published a paper that is seen as one of the first sketches of an idea known as determinism. He believed that if someone could know everything about the universe at any moment in time – all the forces at work, the positions of all the objects – with perfect accuracy, then they could, in theory, wind the clock forward and have perfect knowledge of the future.

Postulating what someone might do if they possessed such knowledge, he wrote, 'they would embrace in a single formula the movements of the greatest bodies of the universe and those of the tiniest atom; for such an intellect nothing would be uncertain and the future just like the past would be present before its eyes.'[3]

Laplace's influence extended well beyond physics. Born in 1749, he was a contemporary of Napoleon Bonaparte. The two men inhabited distinct realms – war and science – yet Laplace's deterministic philosophy resonated with Napoleon's desire for control and strategic precision. Napoleon yearned to understand the battlefield like Laplace understood the cosmos, meticulously calculating troop movements, leveraging terrain, and even employing astronomers for precise artillery targeting. Napoleon envisioned armies as celestial bodies, their movements governed by gravitational pulls of strategy, morale and logistics. He meticulously balanced offensive and defensive forces, seeking advantageous points of leverage like fulcrums, and exploiting enemy weaknesses like gravitational slingshots. This physics-infused world-view extended beyond warfare.

---

[3] *A Philosophical Essay on Probabilities,* Pierre-Simon Laplace. Translated from the sixth French edition by Frederick Wilson Truscott and Frederick Lincoln Emory. New York: Dover Publications, 1951. Originally published in 1814.

Laplace's emphasis on laws and order resonated with Napoleon's desire for a stable, centralised France. He sought to impose a rational, scientific structure on French society, mirroring the predictable movements of celestial bodies.

In addition to Napoleon, Laplace's ideas greatly influenced other non-scientists throughout the 19th century, shaping many aspects of modern life. The French philosopher Auguste Comte, for example, came to believe that society could be understood using the same scientific methods applied to the natural world – an idea that would later lead to the development of modern sociology. The Belgian statistician and social scientist Adolphe Quetelet applied Laplace's methods to the study of human behaviour, which would ultimately lead to the development of the field of criminology. Laplace's deterministic outlook even found its way into literature, with great writers like Thomas Hardy and Leo Tolstoy exploring themes of determinism and the limits of human control in their novels. Laplace's work was of course not without its critics. The philosopher Henri Bergson argued that Laplace's determinism negated the possibility of human free will. Others, like the novelist Fyodor Dostoevsky, found his mechanistic world-view to be cold and unfeeling.

Yet despite these reservations, I was hooked. My once antagonistic relationship with physics began blossoming into a love affair. However impossible, I wanted to live in Laplace's deterministic world. For the first time in my life, I had a reason to work hard. I gained a little momentum. My grades slowly improved and much to my teachers' and parents' surprise, I chose to study physics at A level and was eventually lucky enough to win a place at Oxford to read physics. I soon started to learn things that Laplace didn't know.

There is chaos theory, which says that some deterministic systems are so sensitive to the tiniest variations in their initial starting conditions that while the outcome may still follow the patterns laid down by physics, actually predicting them is impossible. And there is quantum physics, based on the uncertainty principle, which says that it's impossible, by definition, to precisely measure both a particle's position and velocity at the same time. The very act of measurement changes the outcome; either its position or its velocity will be affected. But I also learned that for over a century, physicists had been continuing the spirit of Laplace's work through the search for a single unified theory that can take into account the unpredictability of chaos theory and the uncertainty of quantum physics, and still explain and link together all aspects of the universe in one all-encompassing model. A simple elegant equation that might fit on a page – the so-called 'theory of everything'.

One of the most intriguing historical characters I came across during my time at Oxford was the little-known German physicist and mathematician Theodor Kaluza. A modest man, Kaluza spoke 17 languages and supposedly taught himself to swim in his thirties just by reading a book. During the summer of 1919, Kaluza sent dozens of letters to Albert Einstein describing what he thought, at the time, might be a potential breakthrough. A few years earlier, Einstein had published his general theory of relativity, a new model for explaining gravity and the behaviour of the universe at epic scales. Einstein's theory was not, however, compatible with the other great pillar of modern physics – quantum theory – which explains the behaviour of the universe at exceedingly small scales. Motivated by a deep desire to unite these theories of the very large and the very small, Kaluza wrote to Einstein with a proposal to add an extra dimension – a fifth dimension – to Einstein's previously four-dimensional model of the universe. Ultimately, Kaluza wasn't able to fully explain both gravity and quantum phenomena in a single theory, but his insight is credited as being a precursor to modern-day superstring theory – our best, at present, hope for developing a theory of everything. More than a century later, finding a single, ultimate theory of the physical universe remains one of the great unsolved problems of science.

As my time at Oxford drew to an end, I began to feel that this goal of unification may forever remain beyond our grasp. And even if physicists did succeed in achieving a theory of everything, it wouldn't be the theory of truly everything, at least not the kind I longed for as a teenage boy. It wouldn't be able to explain all of life's complexities. And so when I finally graduated, I made the tough decision to leave the world of physics behind, embarking on a very different path. I became a management consultant, and then a policy adviser and speech-writer in the British government, and then a strategy director in the technology industry. None of these jobs had anything directly to do with physics. And so as the years passed, my old university textbooks slowly gathered dust in the attic. I soon forgot how to carry out the advanced calculus needed to model the behaviour of subatomic particles, or the differential geometry that underpins general relativity, or any number of other complex topics I had once laboured to master as a student. But I never stopped thinking about physics and found myself continuing to apply the skills I had learned in situations far removed from physics. In fact, whether writing government policies or business strategies, the ability to think like a physicist – breaking things down to their simplest form, coming up with and testing hypotheses, questioning results, and then finding answers to challenging problems – proved invaluable. Physics provided a different way of thinking about and looking at the world. Moreover, ideas from physics were able to help me make sense of many of the same complexities that I struggled with as a teenager. It's these ideas I share in this book.

When I have felt buffeted by forces beyond my control, I have for example thought about chaos theory. Many of us have heard about the butterfly effect – the beating of the wings of a hypothetical butterfly in one hemisphere sparking a hurricane in the next. Physics teaches us that this can happen because even in seemingly deterministic systems, the minutest variation at the start of something can have a dramatic influence on the final outcome, making it almost impossible to predict what will happen. And this principle applies just as well to the catastrophes that hit economies and national destinies. This was a hard lesson I learned early in my career when the dot-com bubble burst, sending financial shock waves around the globe and precipitating a chain of events that ultimately led to me, thousands of miles away and months later, losing my job. In 'The Physics of Getting Fired' we will see how simple changes can quickly lead to deep, intricate complexity. One little tweak, and events spiral out of control. The good news is that physicists have used this knowledge to develop new ways of thinking about how to predict the unpredictable. Catastrophes cannot be prevented – but we can adjust our thinking to prepare for them.

When I felt confused by the idiosyncrasies of human decision-making, I turned to quantum physics. Very few people will admit to making irrational decisions that fly in the face of logic and reason. This leads you to wonder why so many people (including you and me) do this, often without any sense of irony or contradiction. Quantum physics is a whole science based on the seemingly counter-intuitive, with hypothetical cats in a box that are both dead and alive, and objects that manage to be both waves and particles, until you look at them. The key to understanding how two seemingly contradictory things can be true at the same time is the principle of superposition. Superposition can apply both to subatomic particles and ideas in the heads of human beings. In 'The Physics of Irrational Decisions' we apply the principles of quantum physics to human cognition to understand irrationality that makes sense. People don't actually think in neat, straight lines – a realisation that makes the irrational far easier to deal with.

When I've found myself feeling that the world is desperately unfair, again, I found myself trying to make sense of this through the eyes of a physicist. In many ways, the universe has always been and will always be unequal. Less than a billionth of a second after the Big Bang, in the vacuum of nothingness, tiny quantum fluctuations in the composition of the universe appeared. Some areas had more energy than others. This inequality allowed pockets of energy and matter to coalesce, eventually leading to the formation of stars and galaxies and planets and ultimately us. In 'The Physics of Having Less' we will see that life as we know it would not have come to exist without inequality; in fact, without it there would be no such thing as existence, and now that we do exist, inequality is simply the most efficient way systems find of distributing themselves. When inequality is eradicated, things cease to function. So, instead of worrying about inequality – or, even more futile, trying to stamp it out – it is far more productive to tackle unfairness. Inequality is ever-present in the universe; unfairness does not have to be.

Sometimes I don't feel confused or buffeted. I feel I am just steadily, resolutely, step by step moving . . . absolutely nowhere. Everything is lined up, there are no obvious obstacles and I am working hard, putting in the hours every day, all for no obvious result at all. My goals are no nearer. In fact, while momentum is associated with movement and inertia with staying still, they are both part of the same thing. In 'The Physics of Going Nowhere', we see that the laws of thermodynamics tell us energy for change cannot just be willed up. More energy in one place means less in another. The energy required for change is often overlooked, and it comes in two kinds: put simply, the useful and the non-useful kind. Energy can be busily pumped out, but if it is the non-useful kind then it is just wasted.

I've even turned to physics when I've had my heart broken. Before meeting my wife, I had a handful of romantic relationships. Some were good. Others felt one misplaced word away from a blazing row. I struggled to recognise the factors that separated the good relationships from the bad ones. As a physicist, I concluded that it all comes down to stability. Physics teaches us the difference between stable systems, where equilibrium is quickly restored after an upset, and unstable systems, where just the slightest nudge can lead to catastrophe. 'The Physics of Breaking Up' offers words of comfort for the recently or frequently heartbroken. And the same principles that apply to romance can also apply to other kinds of relationships, from friends and family to the relationship between governments and volatile voters, leading to outcomes like the UK's break-up from the European Union after Brexit. Physics also gives us the counter-intuitive notion that sometimes the obvious course has the opposite effect desired.

Since getting married and moving up to the north of England, I haven't always found it easy to fit in. My own kids make fun of how I pronounce certain words like 'bath' or 'grass'. I can feel like I'm in my own little bubble. Physics tells us how bubbles act. The laws of fluid dynamics mean that their boundaries move at angles to obstacles, they make themselves as short as they can, they merge to form bigger ones, and eventually they settle down into equilibrium. In 'The Physics of Not Fitting In' we will see that the boundaries of language ebb and flow in the same way. The image of bubbles gives us a vocabulary for understanding the constant churn of social forces and the behaviour of cultural trends.

And while the forces that shape our lives are sometimes invisible and unquantifiable, their effect can be felt as strongly as gravity or magnetism. This was a lesson I learned in my early forties, a time when I went through the most clichéd of midlife crises. I felt inadequate and hollow, questioning so many of my choices. In fact, around the time I was grappling with this minor personal crisis, the world faced its own, with the start of the Covid pandemic. In 'The Physics of a Midlife Crisis' we see that our behaviour and attitudes, consciously or unconsciously, are a response to forces around us. Patterns of behaviour, ideas and feelings spread through communities based on what those nearest to us are doing.

* * *

Before I begin, it's worth clarifying a few important points. I am almost certain that Isaac Newton was not thinking about climate change or global poverty – let alone my career or love life – when he developed his theorems hundreds of years ago. His work, and the work of countless other great physicists since, was about explaining the physical universe, not the human one. So it's natural to ask, is there any real connection between physics and the world beyond physics? Or are the parallels I'm drawing just coincidences?

Putting the specific subject of physics to one side for the moment, I am a great believer in the general power of lateral thinking. This term, first coined by Maltese physician and psychologist Edward de Bono in 1957, is about trying to solve seemingly intractable problems indirectly. Sometimes it is necessary to tackle problems from an outside perspective in order to shift thinking. I have seen first-hand the virtues of such lateral thinking. I've learned that it's important to look for ideas in as many different places as possible. And so sometimes, the value of physics lies not in the concrete or quantifiable, but rather in the metaphor or analogy. It is about thinking in a different way.

But why physics, as opposed to any other science? If the adolescent Zahaan had found meaning in another of my school's science offerings – chemistry, say, or biology – would that be the topic of this book instead? What does physics bring to the table that other sciences do not? There is a glib, but misguided, answer to that question. A philosopher would call it reductionism, which asserts that the secret to understanding any complex system is to break it down into its component parts. If you can understand the parts, the thinking goes, then you can understand the system. A common belief among reductionists is that all science can ultimately be reduced to physics. For example, the foundation of medicine is biology (it's all about making living organisms better), the foundation of biology is chemistry (our bodies are just complex collections of interacting chemicals), and the foundation of chemistry is physics (it's all about the positions of and relations between atoms and molecules). So, all knowledge of medicine can, in theory, be derived from physics.

There are others who take an even stronger view, believing that other aspects of our lives – such as economics, psychology and politics – are also essentially just physical processes that can similarly be broken down into smaller and smaller parts that eventually lead back to physics. This is not my view. I am a reductionist, but not as reductionist as that. There are areas where this kind of extreme approach simply does not make sense. Take politics as an example. An extreme reductionist would say this too comes down to physics, because politics is about the behaviour of groups of people; groups are made of individual people, people are made of atoms, and atoms are governed by physics. But does this mean we could predict the outcome of, say, a national election by studying the behaviour of individual atoms inside the brains of every registered voter? Unlikely. In the first instance, we know that we can never have a perfectly accurate picture of every atom in every brain; the laws of quantum physics tell us that there will always be uncertainty. And even if it were theoretically possible, what would be the point? Theorising is not the preserve of physics. Economists already have economics. Psychologists already have psychology. Politicians already have political theory. Physics might contribute insights to these successful disciplines, but it can't replace them.

For me, physics is about the striving to understand systems in as fundamental and general a way as possible. And in recent years a number of scientists have found a way to use physics to tackle real-world challenges without needing to resort to extreme reductionism. For this book, I may not have interviewed Dominic Cummings or Elon Musk, but I have spoken to a physicist applying the principles of thermodynamics to combat our insatiable greed for fossil fuels. I have met a physicist who has found a connection between flow systems and the seemingly unstoppable growth in wealth inequality. I have talked to a physicist using the theory of how water evaporates into steam – a process known as phase transitions – to predict and prevent volatile election outcomes. I have met a physicist applying theories of quantum physics to better predict and explain irrational and biased decision-making. And I have even come across someone who has been able to use techniques from statistical physics to predict riots and social violence a decade before they actually happen. All of these people and stories appear in this book.

And so, inspired by Douglas Adams and his own protagonist's journey through an imperfect universe, let's begin the tour of our mysterious and chaotic galaxy, with physics as our guide.

Chapter 1

THE PHYSICS OF GETTING FIRED

During my final year at Oxford, the hot career for aspiring young graduates was a job in management consultancy. Entering the university careers service on Banbury Road, it felt like every other shiny brochure was for an elite consulting firm, promising intellectual stimulation, glamorous travel and the opportunity to work with interesting companies. It all sounded great. But I had no idea what management consultants actually *did*.

There's an old joke that consultants borrow your watch so they can tell you the time – and then charge you for it. After a bit of research, I discovered that management consultants advise companies and other organisations on how to tackle any number of strategic and operational challenges that they themselves don't have the time, experience or perspective to solve.

The marketing brochures explained that the best consultants are exceptionally structured thinkers, able to take a swirl of disconnected information and make sense of it. They are also highly curious, always asking questions and challenging the status quo. In many ways, it felt like the perfect career for a former student of physics.

For centuries, physicists have prided themselves on exactly these skills. Physicists stared at the skies and wondered: 'Why are we here?' 'Why is there something and not nothing?' 'Why did that just happen?' And over countless years, they slowly learned how to harness that curiosity and direct it towards understanding the world around us in a structured and systematic way. From the ancient Greeks to Galileo and his telescope, this process became known as the scientific method, a technique for learning about the world in a systematic, replicable way. To this day, the scientific method remains the most fundamental approach used by physicists – by all scientists, in fact.

There are six basic steps to the scientific method. Step 1: Make an observation. For example, the sky is blue. Step 2: Ask a question. Why is the sky blue? Step 3: Form a hypothesis. The sky is blue because of the scattering of electromagnetic radiation, such as light, by air molecules. Step 4: Make a prediction based on the hypothesis. If scattering is the reason why the sky is blue, then the colour of the sky should change if we observe it at different altitudes and times of day. Step 5: Test the prediction. When we observe the sky at different altitudes and times of day, we see that the sky appears darker blue at higher altitudes and at sunrise and sunset. Step 6. Iterate. We could observe the sky from different locations on Earth, or we could observe the sky from space.

The scientific method provides a structure for revealing the mysteries of the physical world. But I have found that it can also help make sense of the world beyond physics. Consulting companies are fond of setting prospective employees challenging case study interview questions. In one such interview, I remember being asked how I might help a large multinational company that was dealing with a staff retention issue. I knew nothing about the company or their specific issues, but the scientific method provided a useful structure for answering the question. I started with an observation: The human resources team had noticed a recent increase in resignations. Ask a question: Why were so many people leaving? Form a hypothesis: The company's compensation is not competitive. Make a prediction: If we increase employee bonus and overtime pay structures, then fewer resignations will occur. Now test it: Let's say that for the next three months, we will track resignations and a third party will conduct exit interviews. And finally iterate: Communicate the findings and re-evaluate compensation and benefits.

The real world is of course far more complicated than an interview case study. In the example above, one might need to develop a number of hypotheses and test each. Some predictions will be accurate and some false. The accurate predictions help you narrow down on the right solution and eventual course of action. Nevertheless, I found these six simple steps an invaluable guide. And after several of these tough interviews, I eventually landed a job with a consulting firm in London.

Right from the start I was thrown into a gruelling schedule, regularly working late into the night. I was still living with my parents at the time, and my poor mother could never get to sleep until I got home from the office, which was often not until well after midnight. Still, I gave everything to this job. It was my first full-time job in the real world and I felt like I was flying. At my first performance evaluation, a manager told me I had 'partner potential'. The future felt like mine to seize.

Nine months in, I was sent to Italy with my fellow graduate hires for mandatory company training. While there, I got an unexpected call to say that on Monday of the following week, I would need to attend an early morning meeting with a partner back in London to discuss potential 'reductions in force'. I had never heard this expression before, but I was told that it was essentially corporate speak for lay-offs. The firm was going to have to make a certain proportion of its staff redundant and on Monday morning I would be told whether I would be one of them.

Despite being a notorious over-worrier, uncharacteristically, I took the phone call in my stride. Every single one of my colleagues had received a similar call and, at the time, I was convinced it would be one of them that was let go. It surely couldn't affect me. I'd be fine, precisely because I'd been working so hard, putting in so much effort, and getting such good feedback and such good performance reviews.

When Monday morning came, I was called into an office on the first floor of the building, several levels down from where everyone normally worked. Sitting across the table was a very large, very senior American partner with a serious face. On his left was a woman from Human Resources. I'd never spoken to the partner before, but I knew him as the Burger King because he would enjoy a Whopper Meal in his office every day for lunch. He got to the point straight away: 'I'm sorry, Zahaan, your services are no longer required at the firm.'

I don't remember exactly what else he or the HR manager said after that. There was a buzzing in my ears. I felt numb, almost like I was going to faint. I do remember him telling me that I shouldn't even clear my desk. It became apparent that the reason why the meeting had taken place on the first floor was because the company wanted me to take the single flight of stairs down to the ground floor, without seeing or speaking to anyone, then leave and go home. It was as abrupt as that. And that was my first experience of the working world.

* * *

Getting fired is tough for anyone, but I think for me it was particularly painful because the way events had unfolded was in complete opposition to the deterministic world that I longed to live in as a teenage boy — a world that was orderly, predictable and fair. I had worked hard as a consultant because I expected my maximal effort to translate into optimal outcomes. It was hard to make sense of what had happened.

Of course, my employer was not run by soulless automatons who delighted in toying with naive young graduates, only to crush them just as it seemed the world was within their grasp. At least, I hope not. It's more likely that they were guided by the bottom line, and what had happened to me made perfect business sense. There was a broader context behind why I lost my job. And once the dust had settled and the shock had worn off, I could begin to see it.

It was the summer of 2001. The world was yet to change quite as irrevocably as it would in September later that year, but still, all was not well. The dot-com bubble had burst a year earlier, sending financial shock waves around the globe. This came at the end of a period of overconfidence and turbocharged investment in just about any company with a dot-com website – then a new, glittering and sexy concept. But it had failed quite spectacularly to pay off. Both investors and entrepreneurs found out the hard way that a dot-com URL alone was not sufficient to guarantee a company's performance. It turned out that more fundamental factors – such as a viable business plan, adequate capital, sufficient financial returns to cover costs and, above all, basic competence of the management – still had a part to play.

It was not just the dot-com companies themselves that went bust. The shock was passed on to related businesses – companies in the consulting, banking and tech sectors that supported them, and which suddenly found themselves with fewer paying customers. The dot-coms were ground zero for the crisis, but the waves radiated outwards throughout the world of finance. Belts had to be tightened, budgets had to be cut, and workforces had to be trimmed.

The harsh HR rule in such circumstances tends to be last in, first out. I was not the first young graduate to be fired abruptly. Nor was I the last. With hindsight, and possibly even some foresight, I should have seen it coming. With a little more detachment I would have been aware of the bigger picture. That, after all, is one reason management consultants are hired in the first place. Detachment, however, is sometimes hard to find. It can be because you, the main character of your story, are simply too immersed in your own narrative. It can be a wilful blindness, or an inability to conceive that things aren't working out as you would prefer they would. The fact is that, time and time again, crises like that seem to build up unnoticed, until suddenly they burst, to the surprise of everyone. The dot-com bust was not the first nor the last time this happened.

In 2008, there was the subprime mortgage crisis. In the United States, irresponsible lending on risky mortgages to those who simply couldn't afford it had led to a housing bubble, which finally and inevitably burst. It began with poor homeowners in a few key states but the shock was felt all around the world.

In 2011, several members of the eurozone – those countries that use the euro as their principal currency – found themselves unable to repay their government debt. One government going broke is bad enough; five in quick succession, all from one of the wealthiest areas of the world, is calamitous. The countries had been accumulating debt for a variety of reasons, and the expected inward flow of capital that would let them pay this off had dried up. Each country would have had more latitude to handle the crisis on their own terms if they still had their own individual currency, but being pegged to one shared currency both severely limited their options, and increased the knock–on effect into other countries.

In 2013, the US Federal Reserve had been spending $85 billion a month to buy US Treasury bonds. This policy, known as quantitative easing, was initiated after the subprime mortgage crisis with the goal of helping to keep US interest rates low. This prompted investors to seek higher returns beyond the US, especially in emerging markets like India, helping to bolster the strength of the rupee. But when, in May 2013, the then chairman of the Federal Reserve, Ben Bernanke, hinted that the US government would start to taper its bond-buying programme, India suddenly found its inward capital flows drying up. Three months later, the rupee hit an all-time low against the US dollar.

In 2015, turbulence on the Chinese stock market was making waves around the world. About 30 per cent was knocked off the price of Chinese shares over a three-week period. So many investors were selling their shares that hundreds of Chinese companies had to suspend their share dealings. Paradoxically, just a few weeks earlier, Chinese stock markets had soared to a seven-year peak as borrowing costs fell and private investors piled in – and therein lay the problem. The stock market high was driven more by momentum than by the actual value of the market and there was nothing to keep it at its peak.

Those private investors were borrowing heavily to raise money, and as the value of stocks (inevitably) fell, the lenders were requiring more cash or other collateral in return.

Then 2017 saw India's attempts at financial reform score a significant own goal for the country's domestic economy, with serious knock-on effects around the globe once again. Still reeling from the after-effects of 2013, the Indian government announced out of the blue that all 500 and 1,000 rupee notes would cease to be legal tender. Put another way, 86 per cent of all currency then in circulation was suddenly declared null and void. The stated goals of this action were sound, and included forcing tax-dodging cash hoarders out into the open, eliminating fake banknotes in circulation, and helping the country prepare for digital currency. However, the rural economy, which relied primarily on cash, was wrecked overnight. Prices of staples like onions, tomatoes and potatoes halved compared to what they had been a year earlier. Profits plunged, debt grew and, as we should now be becoming accustomed to, a slump in one sector led to down-turns elsewhere. In the first few months of 2017 around 1.5 million people lost their jobs.

The point here is not to have fun highlighting how clever financial schemes have backfired, or how experts missed what hindsight shows to have been obvious. The point is rather to show how sometimes just one small failure in one small area can spark off a cascade of failure across something much larger, leading to markets crashing, businesses crumbling and hopeful young graduates losing their jobs. Can physics help make sense of what's happening? To answer this question we have to go back to the very first physics I learned as a teenager.

* * *

Isaac Newton's laws of motion provided the foundations for modern physics. They proved immensely successful in predicting and explaining the behaviour of objects in our everyday world. If force A is applied to object B then we can know precisely the resulting effect C. They represented a deterministic universe that is predictable and certain. However, this certainty was not to last.

Physicists began to realise that making such predictions was not always so straightforward. Imagine two objects, say a single planet orbiting a star. Using Newton's laws, we can precisely model the behaviour of both the planet and the star over time, pretty much indefinitely. Now imagine introducing a third object, say a moon. You might think, no big deal. It's only three objects; surely the same laws of physics that can launch rockets into space could still predict the motion and behaviour of all three?

It turns out they can't. The addition of this third object leads to immense complexity – so much complexity, in fact, that making predictions over time about the gravitational interactions between the objects becomes exceedingly hard, if not impossible. Imagine all three objects starting in known, fixed positions. After a certain time they will all move and end up in new positions. Now imagine doing the exact same thing again, repositioning the same objects in the same starting positions. After the same time has elapsed, you might expect the objects to end up in the same place as before, but in reality they could be somewhere very different. You could repeat this over and over again, each time starting in the same position, but each time seeing different end positions.

During the late 19th and early 20th centuries, French mathematician Henri Poincaré took on this so-called 'three-body problem'. His groundbreaking work revealed why making predictions was so hard. It's because even in deterministic systems, the tiniest perturbations in initial conditions can lead to vastly different outcomes over time. It's the same phenomenon that years later would become known as the 'butterfly effect'; a butterfly that flaps its wings in China could cause a hurricane in Texas, a metaphor for how a minuscule change here can set off a chain of events that leads to a dramatic impact there.

Poincaré's work represented a turning point for physics. It forced a shift in focus from deterministic, well-ordered systems to those characterised by complexity and chaos. Modern physicists have sought to develop models that illustrate this phenomenon, providing a bridge between Newtonian predictability and the inherent instability of complex systems.

One such individual was the Danish physicist Per Bak. In the 1980s, he was employed at Brookhaven National Laboratory, a US Department of Energy facility based in upstate New York. His research focused on the physics of complex systems and how small events can trigger major events. In 1987, he teamed up with two postdoctoral researchers, Chao Tang and Kurt Wiesenfeld, to publish a paper that would revolutionise the thinking about such phenomena. Their idea was called self-organised criticality,[1] and their surprising inspiration was grains of sand. To illustrate his ideas, let's consider two simple experiments involving sand.

Imagine going to the beach and collecting a handful of sand. The sand is dry and you're able to very carefully pick out individual grains with a pair of tweezers and then weigh

---

[1] 'Self-organized criticality', Per Bak, Chao Tang, and Kurt Wiesenfeld, *Physical Review* A, 38, 364, 1 July 1988. https://journals.aps.org/pra/abstract/10.1103/PhysRevA.38.364.

them, one by one. It would, of course, be a very tedious process, but if you had an adequately sensitive set of scales and were able to individually weigh, say, 500 individual grains of sand, what results would you expect? If you were to plot a graph showing the various weights recorded on the horizontal axis and the frequency of that weight being recorded on the vertical axis, you would almost certainly see a peak in the middle with the curve gradually tapering off on either side. In other words, most grains of sand would have a mass at or around the mean (which happens to be about 150 micrograms) while some would be heavier and some lighter. The heavier or lighter the grain of sand, the less common they will likely become.

This shape is known as a bell curve, or a 'normal distribution', and it's incredibly common in society and nature. Whether you're measuring the average birth weight of babies in a hospital or scores in standardised tests or the average number of people taking the Tube on any given Monday in London or the sizes of leaves on a tree, you will invariably find a bell-curve distribution. As long as the things being measured are independent and not interacting with each other, you will always see the same shape.

Now let's take our grains of sand and do a different experiment. Picture a large table in the centre of a room. Imagine dropping grains of sand onto the centre of the table. One by one the grains fall and a larger and larger pile will slowly build up. Eventually the pile will get so big that there will be a small 'avalanche' and some portion of the sand will slide down the pile. There may be several of these small avalanches, but if you keep dropping grains of sand on top of the pile you may eventually see a much larger avalanche where a significant portion of the pile will collapse.

If you were to draw another graph, plotting the size of avalanches on one axis and their frequency on another, what shape would you expect? You may expect another bell-curve, but you would be wrong. The actual graph would more closely resemble a children's slide at a play park. Starting from the left, you would see a high peak representing the small, frequent avalanches. Moving to the right, this peak would rapidly decline, forming a long flat tail extending far to the right. What such a graph tells us is that the vast majority of avalanches would be very small, but on rare occasions, you would see an extremely large one.

This curve is known as a 'long tail' and this pattern is also common in nature, although in very different circumstances. The severity and frequency, for instance, of forest fires, earthquakes, floods and epidemics all follow similar long-tail patterns. In other words, most are small, but very rarely they can be catastrophically large. Wars also follow a long-tail distribution. Most are short-lived and have relatively few casualties, but very rarely, wars can lead to the deaths of millions.

These two shapes – the bell curve and the long tail – are all around us and it's easy to take them for granted. But they describe two very different things. Scott E. Page, a Professor of Complexity, Social Science and Management at the University of Michigan compares the two by considering the average height of human beings around the world. This currently looks like a bell curve with the mean height of a grown man globally being around 5 foot 9 inches. There are some shorter men (like me) and some taller men, but you very rarely see men shorter than 3 foot or taller than 7 foot. Page imagines what would happen if the mean height of men remained 5 foot 9 inches, but the distribution followed a long tail: 60,000 people would be over 9 foot tall, 10,000 people would be over 17 foot tall, and one person would be over 1,000 foot tall. And to balance out all these tall people, 170 million men would be a Lilliputian 7 inches tall.

With such dramatic differences between bell-curve and long-tail graphs, the next logical question is why? Why do we sometimes see a bell curve and sometimes see a long tail? Is there a model that can help explain this? This was exactly the question asked by Bak, Tang and Wiesenfeld. And to find their answer, they delved deeper into the behaviour of sandpiles.

Consider the same table as before, with the pile of sand on it, but this time the surface is covered with a large grid, much like the squares on a chessboard. Now imagine dropping individual grains of sand onto the centre of the table. For the purpose of this model, each square on the grid can only hold a limited number of grains of sand, a maximum of just three. Once a fourth is added, the square becomes 'full'. At this point, there's a little avalanche and a grain of sand falls into each of the neighbouring squares. One grain goes to the square above, one goes below, one goes to the left and one goes to the right. As the grains of sand continue to fall, more and more squares will get loaded up with grains of sand. At the start, an avalanche might travel only two or three squares and then stop. But when every cell is loaded, one extra tiny grain of sand could send a larger wave, an avalanche, across the whole board.

Bak, Tang and Wiesenfeld were not able to carry out this experiment using actual sand. It turns out even dry sand is too sticky. So they had to make do with a computer model that built up a virtual pile, grain by grain, and coloured in the piles for them. Relatively flat and stable areas were coloured green, and the steeper sections were coloured red.

As you would expect, the pile was at first shaded green throughout. As it grew, red areas began to emerge. Some of the red piles developed on their own, isolated from others. In other cases, a cluster of red piles developed next to each other. When a grain of sand landed on an isolated pile, the avalanche was likely to be a one-off and the pile would quickly return to green. If, however, there were other red areas nearby, then an additional grain could set off a chain reaction of further avalanches in other red areas.

Crucially though, they found that predicting the actual size or precise location of the next avalanche was impossible. The pattern of avalanches was not neat or orderly. Ripples spreading out from the centre of the pile could head off in any direction. Grains of sand seemed to tumble at random. Sometimes nothing happened, sometimes a few other grains tumbled beside it, and sometimes a fresh avalanche would be triggered. Bak, Tang and Wiesenfeld could not say which grain would cause that avalanche, and they could not say how large the avalanche would be. Avalanches had nothing to do with the size of the sandpile or the number of grains. Identical inputs and conditions led to a different outcome every time.

Jordan Ellenberg, a professor of mathematics at the University of Wisconsin-Madison, went so far as to describe a sandpile as almost seeming to be alive. The grains of sand seem to have an agency of their own, speeding up and slowing down as if being acted upon by some external force. The only pattern was that there was no pattern.

Yet, despite this seeming unpredictability, an order of a sort did emerge. Bak, Tang and Wiesenfeld paid particular attention to the respective sizes of the avalanches. Most were small. Fewer were mid-sized. And a rare number were extreme. In other words, the distribution of avalanche sizes followed the long tail.

The power and beauty of Bak, Tang and Wiesenfeld's sandpile model was that it revealed the underlying processes that led to other long-tail distributions. Whether dealing with earthquakes or forest fires or epidemics, the same thing is happening. A system operates as it should. But eventually, a point known as self-organised criticality is reached. It's at this moment when just the smallest change – an apparently insignificant event like one extra grain of sand – can lead to an avalanche, the size and consequences of which are unpredictable. The system could absorb the avalanche and return to stability. Or it could lead to a catastrophic cascade of further avalanches. Complex systems self-organise and build up the potential for their own destruction.

Dr Ted Lewis, a former executive director of the Center for Homeland Defense and Security, cites the 2008 financial meltdown mentioned earlier as a classic example of this sandpile phenomenon. It was years in the making as homeowners ended up owing more money on their homes than they were worth. In the US housing market, the carrying capacity, or the maximum risk the market could bear, is normally approximately 62 per cent. By the year 2008, this figure was 65 per cent, representing an enormous tension. The critical point was exceeded and the collapse occurred without warning. It just took one small incident – a failure in one savings and loan company in southern California, which normally would have passed unnoticed by almost everyone – to precipitate a major avalanche.

Some scientists see sandpiles as a way of explaining complex biological behaviour. For example, if you have ever watched a murmuration of starlings – one of the wonders of nature – then you have seen it in action. A cloud of thousands of birds, wheeling together in perfect formation, twisting and writhing through the sky and forming the most fantastic abstract shapes. Yet, the birds are not communicating in any way. There is nothing calculated or pre-planned, and it would be impossible to predict the flight pattern of any one starling. No one starling would be able to take control and direct the outcome. Every murmuration is unique, but the mathematical rules behind it are very simple. The shifting cloud of birds is like a sandpile avalanche – unpredictable and only occurring when a critical point has been exceeded, followed by a subsequent build-up of tension that once again reaches the critical point and another avalanche (or direction change) occurs.

Even political movements are subject to the sandpile effect. For instance, on 4 January 2011 a street vendor in Egypt of whom hardly anyone had ever heard, Tarek el-Tayeb Mohamed Bouazizi, publicly doused himself with petrol on a busy street and set himself alight following a run-in with corrupt authorities. Normally this wouldn't have left a ripple in world affairs. However, at the time this happened, resentment against the totalitarian behaviour of many authoritarian regimes in North Africa had built up to the point where Bouazizi's self-immolation led to massive upheaval across the region. This was the Arab Spring. If Bouazizi had taken his own life, say, one year earlier, then probably no one beyond his family and friends would have been affected by his death.

The sandpile model reveals that long-tail distributions are a feature of organisations that are interconnected and interdependent. Tension builds up until self-organised criticality is reached and, at this point, the smallest change can lead to much larger cascades. So is there a way to prevent such catastrophes?

For all the recently redundant graduates out there – or those with more foresight than me – the sandpile model offers little comfort when trying to protect specific job losses. Individual employees are best compared to the individual grains, buffeted by shifting sandpiles beyond their control. We can never know or predict the size or location of the next avalanche.

Companies and markets on the other hand, more akin to the larger piles of sand, do have more agency. They can take preventative measures simply by the way they operate. One might think that these measures would involve somehow working smarter – optimising and improving efficiency to prevent future collapses. But in fact the opposite is true. The best way to prevent catastrophes is to prevent the build-up of risk in the first place. And Bak's radical idea was that the way to do that was to become *less* efficient.

Consider a power network. If every household in a city drew exactly the same electrical current, all the time, then the electrical system would be much simpler. Demand would always be predictable and there would be very few outages. This, however, is rarely the case. Perhaps everyone switches on the air conditioning in hot weather. Perhaps millions of people switch on the kettle during the World Cup half-time interval. Or perhaps lightning hits a mains box. In any of these cases, and many others besides, there will be a surge of power, and, if the system does not have inbuilt protection, then across the land, fuses will trip. In this case, the power grid may fail due to inadequate transmission capacity, inadequate fuel for generators, or simply a broken connection. Like grains of sand falling onto a pile, an avalanche of collapsing power may spread like an epidemic throughout the entire grid.

The Lebanese-American essayist and mathematical statistician Nassim Taleb describes events like these as 'black swans'. Swans across most of the world are white. The only native population of black swans is in Australia. So, until the precise moment when European explorers reached Australia and saw a black swan, there could have been no inkling outside Australia that such birds existed. The first explorer to see one could have had no way of guessing that a black swan lay just around the corner – until they saw it. In modern parlance, a black swan is something significant that could happen, but we have no idea of knowing where, or when, or how, or even if it actually will. Until it does.

Black swan events occur far more widely than just in power systems. Even though these unforeseen events can't by definition be pinpointed in advance, they can be catered for. And the counter-intuitive answer is to actually reduce efficiency – to optimise by not optimising – leaving enough slack or redundancy within a system to deal with the unexpected event. A power system can be quite easily protected against surges. One way would be to have plenty of extra cabling and fuse protection, which most of the time will sit idle until the surge strikes. Another way would be to run the entire power network at suboptimal performance, always leaving a little extra capacity for the surges, instead of maximising short-term profit by running it as tightly and within as closely defined parameters as possible.

Such solutions are obviously easier said than done. We live in a just-in-time world. Modern organisations strive to increase their efficiency and optimise their operations by sweating their assets. Reducing efficiency is not a popular solution with politicians, business people or accountants. Shareholders do not like being told that there is spare, unused capacity in their system, or that money has been spent on installing expensive components that often sit unused. As a result, investment in surge capacity is often not prioritised and this can actually make us more vulnerable to catastrophes. And given the catastrophic nature of such black swan events, when the surge capacity is needed, it is really needed.

There is another argument against such preventative measures: black swan events may not always be undesirable. It is true that in an area like homeland defence you very much do not want them. However, innovators like Steve Jobs, Bill Gates or Jeff Bezos could very well be seen as individual black swans, mavericks against the consensus norm way of operating in business, and their actions drive the economy. Picking which kind of black swan to defend against is problematic, because by their very nature they are unforeseen and cannot be planned for.

Even when plans are made, they cannot always be perfect. In a 2011 interview, Lewis said that he was confident about the American political system because it was already deliberately suboptimal. The Founding Fathers wanted it to be inefficient; they wanted the American government to be immune from the same kind of totalitarian tendencies they discerned in George III. Therefore, they built in a system of division of powers and checks and balances, defined by the US Constitution. And indeed, for 200 years more, the Constitution survived every black swan thrown at it, including a civil war. Then, in January 2021, a mob that refused to accept the result of a US election stormed the Capitol to prevent the ratification of that election – a process laid down in the same Constitution. A new black swan had entered the arena, one we will talk about in a later chapter.

* * *

So, what advice would I give myself starting my career again? First, consider that anyone who says they know for sure what is coming – even the smartest management consultant – almost certainly does not. We have shown that blithe predictions, or for that matter even extremely well-informed decisions, cannot be made in a complex system where the impossibility of accurate prediction is inbuilt. Probability can be assessed, but should not be taken as absolute. There is simply no way to know exactly what markets will do next, or what the reaction will be to whatever they do.

So, listen to more than one individual expert and calculate a path that weaves the best way between all their predicted outcomes. Understand that we are all subject to massive, complex systems, and sometimes, despite our best efforts, events far away will have a bigger impact on our immediate lives than our own actions. Therefore, we need to pay attention to the big picture, stay philosophical and remain flexible. Always have a Plan B.

Second, many companies are so tightly optimised that they effectively live in dread of the next downturn. But while downturns – falling grains of sand – will always come, most will probably not lead to a crash – an avalanche – simply because most downturns do not. Your company does not have to run in a permanent crisis-management mode, but in order to be resilient, it does need to build in some redundancy. Concentrate on those things that are within your control, like the company culture and employee well-being. Learn to value those things that bring value to your firm without necessarily showing up on the balance sheet.

Getting fired hit me like a physical blow; the damage would have been halved if I had at least walked into that room with the possibility at the back of my mind that I might be about to be fired. As it is, part of the impact of sudden redundancy is psychological. Your confidence in yourself and your faith in the goodwill of others take a severe knock, and it can take a long time to pick yourself back up and tackle the job market again. The more surge capacity you have as an individual, the more likely you are to bounce back. Instead of working at 100 per cent of your capacity for your employer, perhaps see if you can do your job equally well at 80 per cent. Use the remaining 20 per cent for emergencies and to look after yourself. Avalanches can bring down a sandpile, but the individual grains survive.

Chapter 2

THE PHYSICS OF
IRRATIONAL DECISIONS

The months that followed losing my job were tough. While all my friends and former colleagues were getting on with their lives and careers, I was unemployed, broke and still living at home with my parents. I was incredibly lucky to have their support, but with the economy continuing to struggle, few companies were taking my calls. More than that, I'd had my confidence shattered. I didn't know what to do next. The future, perhaps more than at any time in my career since, felt uncertain. Fortunately, after months of searching, I had one potential job opportunity – and a decision to make.

After my disillusioning experience working in the private sector, I was keen to do something that I felt had greater social value. I decided that I wanted to go into the public sector and become a government policy adviser.

The UK prime minister at the time, Tony Blair, was riding a wave of immense popularity after his landslide election victory in 1997. His vision for the future was known as 'the third way' – a political view that sought a middle ground, blending social justice with economic liberalism, supposedly combining the best of both left and right ideologies. A central part of this vision at the time was to extend the benefits of the internet to more people.

While many of us now take reliable high-speed internet for granted, back in 2001 its use was far from ubiquitous, outside a very limited range of occupations. The World Wide Web was technically invented in 1989 and released to the world at large in 1993. By 2001, most companies and many households had access in some form, but still it was scarce, expensive and unreliable. Most domestic access was via a dial-up modem box that sat between your computer and a phone line, screeching like a banshee.

Thankfully a far better alternative, faster and generally more convenient, was slowly being rolled out across the United Kingdom: high-speed broadband. To this end, Blair had created a new government unit tasked with stimulating the market for broadband. As with the provision of any new technology, the country needed both more supply and more demand. More supply meant getting telecommunication companies to invest in infrastructure, laying the cables and launching the satellites needed to extend the UK's broadband network. More demand meant convincing businesses and regular people to adopt and pay for broadband. Believe it or not, there was a lot of scepticism about broadband at the time and I would often hear friends and colleagues ask, 'Why would anyone need anything faster than 56 kilobits per second?' or, 'Is anyone really going to watch videos online?'

Faced with this challenge, the new government unit was building a team to help roll out high-speed broadband and extend the social and economic benefits of the nascent World Wide Web. And I was lucky enough to be offered a job as a policy adviser and speech-writer in that unit.

You might think that with only one job offer on the table, the decision about whether to accept it would have been straightforward. But there was something holding me back. The job I had been offered was only an 11-month temporary contract. There were no benefits, the salary was significantly less than I was making before, and there was no guarantee of a full-time job at the end of it. Furthermore, I knew that new government units often had a short lifespan and could be shut down at a moment's notice on the whim of a political party.

As someone drawn to certainty and predictability, this ambiguity worried me. Yet I needed to let the Civil Service know whether I would accept or reject. I couldn't stop asking myself one question: Would I still have a job in a year's time? As an obsessive worrier, I decided to play out the various scenarios in my head. I asked myself first what I would do in the event that the temporary job became permanent. Unsurprisingly, I decided that I would definitely take the job. I then asked myself what I might do if the opposite happened. What decision would I take if I knew that the role would end and that I was going to be unemployed again in under a year? This was a harder question. But, perhaps motivated by my desire just to get back to work and the appeal of doing something with greater social value, I eventually decided that I would actually still take the government role. This was a significant breakthrough in my decision-making. Even though I still did not know what the future would hold, I had decided that I would actually be happy and willing to take the government role regardless of what happened.

So did I go ahead and just take the government job? I could have and should have. But I didn't. Instead, I procrastinated. Days passed and I still did not let them know my decision. I kept delaying. The hiring manager was incredibly patient and understanding – at first. But I could tell that he was getting increasingly frustrated. Eventually, after two weeks of obfuscation, I finally accepted the job.

While the manager was happy, I could tell that, before my first day, I had tarnished my reputation. It was not the best start to this new chapter of my career. One might think that I was just being cautious and that waiting was a prudent move. But I had already decided that I would be happy to take the job in either of these scenarios. So logically, delaying didn't make sense. In the grand scheme of things, it may have been a trivial matter, but it was nonetheless an irrational one. Can physics make sense of what happened?

At the time of my decision, I was in a state of limbo. There were two paths ahead of me and I had to decide which way to go. Up until the moment I made my decision, both paths were a potential reality. Physicists have a name for situations like this. They call it a 'superposition' state – a term used to describe a situation in which something exists in two or more states simultaneously. Imagine a coin spinning in the air. For a quantum physicist, while the coin is spinning, it's not just heads or tails; it exists in a state of superposition where it holds the potential for both heads and tails at the same time.

Perhaps the most well-known example illustrating this concept is Schrödinger's cat. This was not a real cat but rather a hypothetical one. In 1935, the physicist Erwin Schrödinger imagined such a cat locked inside a box containing a sample of radioactive material. This material is made of millions of unstable atomic nuclei, which could at any time 'decay' and spontaneously emit a burst of radiation. If the sample did decay, the radiation would cause a hammer to fall, breaking a vial of deadly poison that would kill the cat. Schrödinger posed the question: Without being able to look inside the box, is the cat dead or alive?

Intuitively, a classical view of the world tells us that the cat must be either dead or alive, we just don't know which. But in the quantum realm, until observed, a particle's state can never be fully precisely known. It's not that we don't know, it's that we *can't* know. Its very nature is indeterminate. And so, Schrödinger argued that until the moment when the box is opened, the cat exists in a state of superposition. It is both dead *and* alive.

Schrödinger's cat is one of the definitive thought experiments of quantum physics and we will come back to it. While I was deciding whether to take the job, I was not facing a vial of deadly poison, like this poor feline, but I was stuck in a metaphorical state of superposition. Of course my decision was almost certainly influenced by emotion, bias and other unseen factors. And I was haunted by one other undeniable fact: We all make bad decisions.

* * *

It's estimated that the average adult makes some 35,000 decisions a day. Most are of course small and innocuous –

one study suggests that over 200 a day are just about what to eat or drink. 'Shall I have coffee or tea?' or, 'Can I resist that chocolate bar and have an apple instead?' Others are more important, affecting our own and others' livelihoods and well-being. 'Should I hire the person with more experience, or the one with the better attitude?' or, 'Should I go to the doctor or wait to see if I get better by myself?' These are decisions that shape our lives in ways both profound and superficial, and are the cause of all our actions.

Whether the impacts are large or small, one fact is certain: Some of our decisions won't make sense to other people, or even to ourselves. A married man confesses to his friend that he had an affair many years ago. Should he come clean with his wife? The friend advises him not to say anything – it was a long time ago and nothing good could come from a confession. The man nods in agreement, then goes straight home and tells his wife the whole story.

Our seemingly irrational decisions also extend to issues of global significance. Imagine two people with differing beliefs about climate change. One person accepts global warming is caused by humans, whereas the other is dismissive of the human role. How might the two react if presented with evidence supporting human-caused global warming? Studies show that the person who accepts the presence of a consensus might strengthen his or her beliefs. But the other person, already distrustful of climate scientists, might view the consensus as confirmation of a conspiracy and 'groupthink'. Both people receive the same evidence but their beliefs change in opposite directions. This phenomenon is known as belief polarisation. Although the impact of any one individual's seemingly irrational decisions on issues like these may be inconsequential, the cumulative impact of millions of people behaving this way can be significant.

For over a century, cognitive psychologists have developed models to explain how and why we make the decisions that we do. In the early 1900s, behaviourism became a dominant trend in psychology, at least in the United States. Behaviourists believed that all our behaviours were learned through interaction with the environment, which meant that our actions were essentially no more than responses to external stimuli. Experts in this area were less interested in studying the mind, because they assumed that internal mental processes couldn't be observed and objectively measured. In fact, they believed that there was little difference between the learning that takes place in humans and that in other animals.

Perhaps the most famous example of this was Pavlov's dogs. During the 1890s, the Russian physiologist Ivan Petrovich Pavlov set up a research project to study salivation in dogs in response to being fed. He predicted that the dogs would salivate when food was placed in front of them. But he also noticed that his dogs would begin to salivate whenever they heard the footsteps of his assistant who was bringing them the food. Pavlov discovered that any object or event that the dogs learned to associate with food would trigger the same response.

From the late 1950s onwards, behaviourism began to be supplanted by the cognitive revolution. Put simply, researchers believed that the human mind could be understood like a computer. The ultimate goal of this intellectual movement was to apply the then nascent fields of artificial intelligence, computer science and neuroscience to the study of how humans acquire knowledge and understanding and make decisions. They believed that humans and computers process information the same way – for example, we take in information, we transform it in some way, we store it and then retrieve it from our memory.

By the early 1980s, the cognitive approach had become the dominant line of research across most branches of psychology, spawning a number of theories that can be described as 'classical' models. They're so commonly and widely applied that today we take them for granted. Classical models of cognition are based on some very simple assumptions, including the sure-thing principle, the commutative axiom and the law of total probability.

All three of these assumptions appear perfectly logical. Yet time and again we see examples of how the 'real' decisions we make defy these rational rules.

The sure-thing principle states that if someone is willing to do something whatever the outcome of a certain event, then they should also be willing to do the same thing even if they don't know the outcome of that event. My experience deciding what job to take was an example of the sure-thing principle. I was happy to accept regardless of the potential long-term uncertainty. My decision should have been a sure thing. Yet I procrastinated, defying the sure-thing principle.

Over the years, several controlled experiments have been carried out, recreating similar situations that illustrate how we all make similarly irrational choices. Imagine for a moment a much simpler decision: whether to bet on a coin toss. Imagine being offered a gamble with the following conditions. If the coin lands on heads, you win £200. If it's tails, you lose £100. Would you play? Let's say that you decided to, and that you're offered two turns if you want them. How would the outcome of the first gamble influence your decision about whether to take a second one?

This two-part gamble has been put to participants in a hypothetical way in several different studies over the years, with three different scenarios presented to them.

A third were told that they'd won their first game. A third were told they'd lost their first game. The remaining third weren't told the outcome of their first game.

Most of the participants – 69 per cent on average – who were told that they'd won chose to play a second time. This makes sense. They'd just pocketed £200 and probably felt confident gambling again. They stood to double their money, and even if they lost the second time, would still have £100 more than when they started. Those who were told that they'd lost the first gamble were a little more cautious, but still more likely to gamble again than not: 59 per cent chose to play a second time, as opposed to the 69 per cent who were told that they'd won. Again, we can understand their thinking. Perhaps in this case, frustrated after the first bet, they wanted to win back their money.

The most interesting results came from those who did not know whether they'd won or lost their first toss. Far fewer of the participants – only 36 per cent – chose to play again. How can this be explained? If most people were happy to go for a second turn whether they'd previously won or lost, then surely they'd also be happy to play if they didn't know the outcome. What difference would it make? Again, it seems to violate the laws of logic, or 'classical decision theory' as psychologists call it.

The second assumption is the commutative axiom, which states that the order of two independent events should not affect their relationship. Many of us will have learned in school that the answer to A + B is identical to B + A. In this case, A and B can literally move or 'commute'; their order in the equation doesn't affect the answer. It feels like common sense and it is also one of the basic assumptions of classical cognition. The order in which you consider two independent questions, A and B, should not matter.

While logically this makes sense, I know from personal experience that this is not always the case. Every day, I pick up my young son from school and, on the drive home, normally ask him a couple of questions: 'Did you have a good day?' and 'Did you eat any vegetables?' In classical models of cognition, the order in which I ask these two questions shouldn't matter. That is, the likelihood of my son saying 'yes' he had a good day and 'yes' he ate his vegetables at lunch should be the same as the likelihood of him giving the same answers if I'd quizzed him in the opposite order.

Personal experience tells me otherwise. If my first question is whether he had a good day, his typical response is 'Yes, it was good.' He will then grudgingly admit to having eaten his greens, too. But if the first question is whether he ate his vegetables, you might picture the sigh I see in the rear-view mirror and the answer would be a considerably less cheerful 'Yes, Daddy, and they were yucky.' Following this answer, a later question about how his day went tends to produce a response that's less than entirely enthusiastic.

To my son, any day that he eats vegetables is by definition a less good day than one on which he didn't, and I've just reminded him of this. In other words, the answer to the first question creates a context that changes the answer to the second question. So the probability of answering 'Yes, I had a good day' and 'Yes, I ate my vegetables' doesn't necessarily equal the probability of answering 'Yes, I ate my vegetables' and 'Yes, I had a good day'. In other words, A + B does not necessarily equal B + A.

Eating vegetables, though it may feel like a catastrophe to my son, is insignificant in a global context. But experiments show that these order effects impact on far more serious issues. In the 1980s, the market research company Gallup ran a poll with the goal of investigating hostility between white and black people. They asked a group of 500 people two sequential yes or no questions: 'Do you think black people dislike white people?' and then 'Do you think white people dislike black people?' They also asked another similarly composed group of 500 people the exact same questions but in the opposite order. Rational logic predicts identical answers across both groups. But in fact those in the second group, who were first asked whether white people dislike black people, were more likely to agree that black people dislike white people but disagree that white people dislike black people.

The final assumption is the law of total probability: The probability of an event happening is equal to the sum of the probabilities of all the mutually exclusive ways that the event can happen. To explain this in a simple way, imagine you're planning a picnic and you want to calculate the probability that it will be ruined by rain. The weather forecast tells you that there is a 70% chance of rain. In order to stay dry, you ask your friend to bring an umbrella. He's an unreliable friend though and is likely only to bring an umbrella 50% of the time. The probability therefore of your picnic being ruined is 70% (it rains) multiplied by 50% (your friend forgets the umbrella), which equals 35%. You can also calculate the probability of your picnic not being ruined by adding 30% (the chance of no rain) plus 35% (the chance of rain but your friend remembering the umbrella), which equals 65%. The law of total probability tells us that when you add the probabilities of all the possible scenarios together (in this case, 35% plus 65%) you must always get 100%.

This sounds obvious enough, but it's something we often get wrong. A famous example is the so-called 'Linda problem', first introduced by the Nobel Prize-winning psychologist Daniel Kahneman in the 1970s. Imagine a hypothetical person called Linda who you are told used to be a philosophy student at a liberal university and active in an anti-nuclear movement. Given this information, which of the following two options is more likely: that Linda is now a bank teller or that Linda is now active in the feminist movement and a bank teller? If you answered the latter, you're not alone. In countless experiments, roughly 85 per cent of people said the same. But you would be wrong. That's because the probability of two events occurring together is always less than or equal to the probability of one of them alone. To think otherwise is irrational.

These three assumptions and others like them sit at the heart of modern cognitive psychology, but as these examples show, they haven't always been able to explain the results that happen in 'real life'. So is there another way of understanding what happened here?

* * *

In recent years, new thinking has emerged to help understand and better predict our seemingly irrational decisions. To explain apparent paradoxes in human decision-making, psychologists and cognitive scientists have found inspiration in another field synonymous with seeming paradoxes, a field that is predicated on uncertainty – where cats can be both dead and alive. Welcome to the world of quantum cognition – the interface between quantum physics and decision-making, and a new field of research that is expanding our understanding of human decision-making.

To learn more, I wanted to talk to one of the world experts in this area. Peter Bruza is a professor at Queensland University of Technology, Australia. He's a computer scientist who has spent years studying how we can better understand the workings of the human mind through quantum theory. His 2012 book *Quantum Models of Cognition and Decision*,[1] which he co-wrote with cognitive scientist Jerome Busemeyer, is the seminal text on the subject.

Our video call takes place on a midweek evening, by his time, and I'm surprised to see him appear on the screen in a Hawaiian shirt. I wonder whether this is how all Australian academics dress, but it turns out that he's just returned from a funeral at which all the guests were asked to wear bright clothes. After checking that he's still okay to talk, I ask him if he can help me to understand more about this area. He launches into our conversation with the enthusiasm typical of a lifelong scientist.

'By its very nature, decision-making involves uncertainty,' he says. 'In our minds we create representations of the external world, and to make decisions about these representations involves computation. That's essentially what cognition is – it's how we understand our experiences and use that understanding to mentally compute

---

[1] *Quantum Models of Cognition and Decision*, Jerome Busemeyer and Peter D. Bruza, Cambridge University Press, 2012.

what choices to make. If you're in a state of certainty at the time of making your decision, you'll come to a different one than if you're in a state of uncertainty.'

This notion of uncertainty is at the core of what it means to be a decision-maker in a complex world. To understand how quantum physics relates to this, we need to enter the realm of particles.

As we saw in the previous chapter, over 300 years ago Isaac Newton developed his groundbreaking laws of motion, describing the relationship between objects and the forces acting on them. For the first time, it was possible to predict how, say, a cricket ball would behave when thrown, how the Moon orbited the Earth, and how a rocket could get there. The laws brilliantly explained the observable universe, and they comprised the model we followed for a long time afterwards – and still do. Newton's laws provided certainty.

But as we also saw in the previous chapter, in the 19th century, scientists began to question this certainty. Even in deterministic systems, minute changes in initial conditions can lead to vastly different outcomes – what is now known as chaos theory. And in the 20th century, scientists would continue to discover that the world we live in is far from certain.

Consider again a cricket ball. You can hold it in your hand and it feels solid, and it is. At any point in time you can, in theory, know where it is or how fast it's going. If you leave the ball under your bed at night, then unless someone or something moves it, you can be certain it will be exactly where you left it in the morning.

Look deeper, though, and the cricket ball is actually made up of billions and billions of subatomic particles – electrons, protons, neutrons – invisible to the naked eye. And the behaviour of these particles is radically different to that of the cricket ball itself. A new theory was needed to explain the behaviour of these subatomic particles, and so was born quantum physics, arguably the most successful scientific theory so far. It explains puzzling findings that were impossible to understand using the older, classical Newtonian theory, and it achieved this by introducing a set of revolutionary principles.

One of these principles was that of uncertainty. Particles – including those contained in the radioactive sample of the Schrödinger's cat thought experiment – are said to exist as waves of probability. Only until the moment of observation (in this case when the box is opened) does this wave 'collapse' to a definite state and we discover whether the cat is either dead or alive.

I remember learning about Schrödinger's cat when I was at school. There was a part of my brain that refused to believe it was actually real. It sounded like an abstract thought experiment that made sense mathematically but had no bearing on reality. Apart from anything else, how can something be both a wave and a particle? Picture my amazement when my physics teacher talked me, step by step, through real-world experimental evidence that showed this is not just a mental trick. This is something that happens.

Without harming any cats, he described a common demonstration that schoolchildren around the world have learned about for over a century: the double-slit experiment.

It starts by imagining a wall with two identical, parallel slits. What would happen if you started throwing shiny red cricket balls, one after the other, at this wall? Many would bounce off, but some would find their way through one or other of the slits. If a second wall was placed behind the first, they would hit it. If each cricket ball collision left a red mark on the second wall, what pattern would you expect to see develop? You'd likely see two strips of red spots emerge, roughly in line with the slits in the first wall.

Now imagine the same experiment, but instead of discrete objects like cricket balls passing through the two slits, it is waves of water. Think of a wave travelling across the surface of a pond. If this wave were to encounter a wall with two slits then it would split into two new waves, each passing through one of the slits and radiating out the other side in concentric circles. If the slits were close enough, the two waves emerging from the slits would collide with each other. A peak meeting a peak would amplify, creating a bigger wave, and a peak meeting a trough would cancel out. This is what's called an interference pattern.

It turns out you always get this same pattern, regardless of what kind of wave passes through the two slits. In 1801, the British polymath Thomas Young shone a beam of light at a wall with two slits and observed the same interference pattern on the other side. This is because light is also a wave. The light passes through both slits, creating two new waves that interfere with each other.

Now let's think about what happens when this same experiment is carried out at a much smaller scale, in the quantum realm. Consider electrons – subatomic particles many orders of magnitude smaller than a cricket ball – being fired at a wall. What would happen? Keeping things simple for the moment, imagine the wall has just one slit, still with a second wall behind it. In this case, the pattern you see on the second wall is similar to that observed with the cricket balls: a long strip similar in size to the slit where the electrons pass through. Particles and cricket balls act in the same way.

Things start to get strange, however, when you introduce the second slit. Intuitively, you expect to see another strip form on the second wall. But that's not what happens. On the second wall you get an interference pattern, similar to that observed when a wave of water passes through the slits.

When this experiment was first carried out in a laboratory, the first assumption was that multiple electrons were somehow interfering with each other as they passed through the slits simultaneously. To test this assumption, scientists slowed down the flow of electrons so that, instead of a barrage of electrons passing through the slits together, literally only one was fired at a time. One electron, then another, then another. Surely, this would prevent any chance of interference? It would be like trying to clap with one hand. Yet incredibly, as each electron was fired, one after the other, the resulting pattern on the second wall eventually revealed . . . an interference pattern.

To get to the bottom of what was going on, scientists placed detectors by the slits. Perhaps the individual electrons were somehow splitting into two, passing through both slits, and then interfering with themselves? Incredibly, as soon as the detectors were put in place, the resulting pattern changed. The interference pattern disappeared and instead became two strips, like we saw with the cricket balls. When the detectors were removed, the interference pattern returned. Somehow, the very act of looking at the electrons changed their behaviour. When the detectors were on, they behaved like cricket balls. When off, the electrons behaved like waves.

Remember that, unlike Schrödinger's cat, this is not a thought experiment. This is an actual experiment that has been tested in a laboratory, and can be repeated by anyone with access to some fairly standard equipment. This simple experiment reveals so much about the beauty and bizarreness of the quantum world. Until the electron is observed, its position can never be precisely known. The electron behaves like a wave of probability, capable of interacting and overlapping with other electrons, much like waves on a pond. At any point, you can estimate where the electron is going to be, but you can never know for certain. Only when detected does the electron have a definite position, acting like a particle.

To the rational mind this outcome doesn't make sense. Why would looking at the particle affect how it behaves? And yet, this is what takes place.

*　*　*

That is the reality of quantum physics. Quantum cognition, mentioned earlier, describes what happens when the principles of quantum theory are applied to cognitive models. It's based on the insight that, just as the quantum universe is predicated on uncertainty, so too is the way we make decisions.

To see how this works, consider again the example of the coin-toss experiment, where the participants' behaviour seemingly violated the sure-thing principle. There are a number of similarities between this and the double-slit experiment. Neither makes sense in the rational, linear world, but they start to be more understandable when we think of uncertainty.

In both experiments, there are two possible paths. For the particle, the two paths are those formed by the two slits. For the coin flipper, the paths are whether the first toss is won or lost.

In both, the path taken can be known or unknown. For the particle, this depends on whether the path is observed. For the coin toss, this depends on whether the participants are told the results of the first gamble.

In both, if the path taken is unknown, a state of super-position is created – in other words, a state where two realities are possible. When the particle is in a state of superposition, it acts like a wave. When the coin flipper is in a state of superposition, there's a lower probability they will choose to gamble again.

And it turns out that when psychologists apply quantum theory to their thinking, they are able to do something really extraordinary: make specific numerical predictions about human decisions. This might sound run-of-the-mill in lots of other sciences, but it is rare in psychology and a challenge that has eluded cognitive science for years. Psychologists usually don't know enough about the underlying causes of people's behaviour to make really accurate predictions.

In his research, Peter Bruza's colleague Jerome Busemeyer has investigated several historic psychological experiments where quantum theory more accurately predicts the real-life results. Let's consider a few of these, starting again with the Gallup poll investigating hostility between white and black people – an example of order effects in the commutative axiom. Half the participants were asked if they think black people dislike white people and then if white people dislike black people. The other half were asked the same questions, but in the opposite order. Classical models of cognition predicted that the answers should have been the same regardless of the order. But quantum cognition predicts a 10 per cent difference – and this was exactly what was observed.

The predictive power of quantum cognition has been applied to many other similar questions. Over a 10-year period, quantum cognition researchers in America looked at 72 Pew surveys spanning a range of political judgements, such as whether you are happy with your president or whether you feel your country is doing well. Each survey had 1,000 participants, with 500 asked each question in one order and the other 500 in the opposite order.

Classical cognition predictions deviated significantly from the observed results, while quantum cognition matched them almost exactly. These predictions were established *a priori*, meaning they were formulated before the events took place. This predictive power is exceedingly rare in social sciences.

While these examples may seem very abstract, quantum cognition can be applied to real-life decisions we all might face. The Linda problem described earlier was an example of how sometimes our inherent biases can lead us to make irrational decisions. Linda's story made it sound like she was likely to be a feminist, leading people to believe that she was more likely now to be a feminist and a bank teller than just a bank teller. In recent years, quantum cognition experts have designed experiments to measure the impact of these biases on the decisions people make, sometimes based on nothing more than appearances.

In one particular experiment run by James Townsend, a professor of psychology at Indiana University, participants were shown pictures of two different types of face: narrow and wide. Purely on the basis of the different face shapes, they were asked to decide on one of two possible courses of action which, for the purposes of this explanation, we can describe as either hiring or firing. The goal of the experiment was to understand which of these two actions was most likely, and under which circumstances. The experiment, however, had a twist. Some of the participants were asked to make a decision straight away. Others were asked to take an extra step – by first categorising the faces as either 'good guys' or 'bad guys' – before then deciding on a course of action.

To make sense of the results, for the sake of simplicity let's just consider what happened when participants were shown pictures of narrow faces. For those who had to decide on a course of action straight away, 69 per cent decided to fire and 31 per cent to hire. It's not clear why a picture of a narrow face moved more people to fire than to hire – but that's not the point of this experiment. The most interesting finding comes when you compare these results with those participants asked to take the extra step: first categorising the faces as either good or bad guys. In this scenario, when shown pictures of narrow faces, the average likelihood of firing, rather than hiring, was only 59 per cent. Adding the extra step reduced the 'firing' rate by 10 points.

If you were to think of the brain as a logical computer, this obviously doesn't make any sense. It should be just as likely that the subjects would decide to fire or hire regardless of whether they made their decision straight away, or whether they categorised the faces as good or bad first. This is what all classical models of cognition predict. And yet we see this significant discrepancy.

How can physics help to make sense of these results? Let's imagine the participants as quantum particles. And let's imagine that instead of travelling through the double slits, they are travelling on a journey from being shown a face to making a decision whether to hire or fire. For those asked to make a decision about hiring or firing straight away, their thought process is the equivalent of being unobserved. They are therefore in a state of superposition. But for those first asked to make a categorisation of good or bad before deciding to fire or hire, their path is the equivalent of being observed. They're no longer in a state of superposition but rather in a definite state. And it turns out that when you apply quantum cognition models to this experiment, it provides a natural explanation for the observed results. Credited to the pioneering work of Busemeyer and Bruza, this research has been a significant breakthrough. Where classical models of cognition break down, quantum cognition explains the real-life experiment results with remarkable accuracy.

*   *   *

While the world of quantum physics might seem abstract and distant from our everyday lives, it holds fascinating insights about how we make decisions. Classical models are limited because they fail to take into account the inherent uncertainty of human decision-making. Quantum cognition allows us to more accurately predict the impact of this human response.

I asked Peter Bruza how he sees quantum cognition helping us make better decisions in the future. Could ideas from quantum physics stop me making irrational decisions? 'One of the big sells of quantum cognition,' he replies, 'is that it gives us some methods to model human thinking, warts and all; because in normal life we tend to make quick decisions without having all the relevant information to hand, and it's the most flexible framework I know for dealing with that.'

Applied to my dilemma about taking the job, I can see now that I was in a metaphorical state of superposition, unsure of which path to take. Quantum cognition teaches us to recognise this state of uncertainty and use it to our advantage. Instead of seeing the decision as a binary choice with a right or wrong answer, I could have acknowledged the probabilities of different outcomes and prepared for multiple scenarios. This means I could have been more comfortable with the inherent uncertainty, reducing my procrastination and making a more timely decision.

If I had known about quantum cognition principles at the time, I might also have developed a strategy that accounted for the possibility of the job not becoming permanent. I might have set specific goals and milestones for the 11-month period, ensuring I gained valuable experience and relationships and in doing so improving my prospects regardless of the job's permanence. This mindset shift could have alleviated my fear of uncertainty and empowered me to act more decisively.

In the quantum realm, particles can exist in multiple states simultaneously – in other words, superposition. Quantum cognition teaches that we too are often in a similar state of superposition when making decisions. We therefore need to avoid making decisions solely based on the perceived certainty of an event. Accounting for the interconnectedness of possibilities fosters a more nuanced understanding of risk and reward. Quantum mechanics posits that the very act of observation impacts the state of a particle. For humans making decisions, this means that simple changes like the order in which questions are asked can lead to different answers.

'The interesting thing is that although we sometimes get things wrong with our irrational decision-making, we don't get them wrong badly enough to kill us off as a species,' Bruza says. 'We've even been able to develop some quite wonderful civilisations. We may be limited cognitively and we may be irrational, but we're certainly on to something.'

As shown by examples like the man trying to decide whether to confess to an affair, humans have an instinctive grasp of quantum cognition without realising it. The moment they do realise the complexity of what they are doing, their thought processes tend to jam up.

Once again, observation affects the outcome. But actively taking quantum cognition on board helps us transcend traditional decision-making models. Incorporating lessons from quantum physics enhances our ability to navigate complexity and ultimately make better decisions.

Chapter 3

THE PHYSICS OF HAVING LESS

A few years ago, a team of researchers at Yale University gathered a group of young children and gave them a simple task. They were asked to divide 10 gold coins between two people, on the basis of their performance on a test. I don't know the two people's names but let's call them Andy and Betty. The children were not told anything about the objective of the experiment. The first time the researchers ran the experiment, the children were told that Andy and Betty had performed equally well on the test. The children thought about it and duly awarded Andy and Betty five coins each. Fair and equal. The second time, Betty did slightly better than Andy. This time the children decided to give Betty six coins and Andy only four. Unequal but still perhaps fair. In the final scenario, the children were told that Andy and Betty had once again scored equally well on the test, but this time they had to distribute 11 coins. Without being allowed to break a coin in half, there was no way to avoid giving Betty and Andy an unequal number of coins. So what did the children do? They simply threw the odd coin away and awarded the remaining 10 equally.

Without any preconception, the children displayed a basic instinct for fairness – a desire either to reward equally or, if they felt it was deserved, to reward unequally but fairly. They had the realisation that things can be unequal but fair, and that unfairness matters more than inequality. The experiment suggests that the ability to make this distinction starts early, and runs deep.

At first glance, equality and fairness are synonymous – but there are important distinctions. A physicist might distinguish the two terms by considering the universe in its earliest moments, which is where these concepts first appeared. Less than a billionth of a second after the Big Bang, the universe was in a state of perfect 'equality' – a singularity of infinite density and temperature where all that was, and all that would be, existed as a perfectly uniform point. Everything was the same everywhere.

Almost immediately, anyone who was around and able to observe might have seen differences start to emerge. Tiny quantum fluctuations emerged within this coherent mass. Some areas of the universe began to have more energy and more heat than others. And as the universe expanded, these pockets of arbitrarily distributed resources became significant, ultimately leading to the formation of stars, galaxies, planets and eventually life itself.

Later in this chapter, we will delve deeper into these quantum fluctuations and the subsequent formation of cosmic structures. But first, let's consider the nature of inequality in the human world. There are many ways it can creep into a society: gender; age; class; disability; ethnicity; religion; sexual orientation; access to technology; access to opportunity. It's a complex mix. But underlying many of these factors is wealth.

Wealth inequality takes various forms. There is certainly desperate poverty in parts of the developing world when you compare them with richer societies, but in recent years there has been a sharp increase in inequality *within* nations. In 1980, inequality within countries made up less than half of global inequality. By 2020, that figure had risen to two-thirds, exacerbated by rapid rises in incomes at the top of many economies, and chronic stagnation for the rest. Between 2013 and 2023, the number of billionaires in the world and their total wealth has doubled, and 52 per cent of the world's income is snapped up by its richest 10 per cent of inhabitants. Meanwhile, in many places, life for ordinary people is getting harder; Oxfam reports that in 2022 at least 1.7 billion workers lived in countries where inflation outstripped wages.

And what is the result? Where there are disparities between higher- and lower-income households – say, in terms of access to education, or healthcare or digital technologies – children of the latter start with a built-in disadvantage that becomes difficult, if not impossible, to ever shake off.

The stark gulf between rich and poor living side by side – including in some of the wealthiest societies on Earth – has changed how we see unfairness, making it more visible and present.

* * *

Over the course of my life, I've had a lot of opportunities and have been incredibly fortunate. I was born in a rich country to a middle-class family. Yet that doesn't mean I haven't occasionally felt the pangs of unfairness and inequality. The period after I left management consultancy was one of these times.

My initial 11-month contract with the Civil Service eventually became permanent and I ended up spending the rest of my twenties working in the public sector. The Civil Service was an excellent employer, and I felt an immense sense of pride and meaning in my work. I also knew that salary-wise, I was doing well relative to most British workers. Yet I was conspicuously aware that there was a big gap between what my colleagues and I earned and the management consultants I still knew. They were now earning six-figure salaries. They were able to afford nicer clothes, cars and apartments. And they were also able to turn money into even more money by investing and buying assets, further widening the gap.

I'm not proud of it, but I felt jealous. And I felt angry. In my view, the work I was doing as a civil servant was just as important, just as challenging, yet somehow I had less. As an insecure young man, it felt unfair. And as I dealt with my own petty emotions, I began to notice that these themes of unfairness and inequality seemed to be ever-present in the world around me. In 2005, I was working for the Department for International Development (DfID), the UK government department whose very mission was to eradicate global poverty. Questions of global fairness and equality were on my mind all day long.

This was never more the case than during three days in July 2005; three days of incredible highs, and lows. Things got off to a very good start. At the 117th Session of the International Olympic Committee in Singapore, London was selected as host city of the 2012 Olympics. Years of bidding and dealing finally paid off as the UK's capital saw off competition from Madrid, Moscow, New York and Paris to become the first city to host the games three times. I watched the announcement with my colleagues on the TV in our offices. As a born-and-bred Londoner, I felt so proud and happy. It was a good example of unequal fairness. Not everyone could be the winner; only one city could host the Games. But, as long as the selection process was clear and unbiased, no one could complain. That was 6 July 2005.

The following morning, I left my flat in the west London neighbourhood of Hammersmith hoping to catch the bus to work. I waited for over 45 minutes at the bus stop, but none came. I eventually gave up waiting and decided to walk. It wasn't until I got to my office desk that I discovered what had happened: Three suicide bombers had detonated explosive devices on the London Underground, and a fourth on a bus in London's Tavistock Square. The attacks claimed the lives of 52 people and injured 700 more. Like most suicide bombings, the question of who was injured and who was not was simply a matter of good or bad fortune. Anyone travelling on the Circle or Piccadilly lines or the number 30 bus route during the morning rush hour on that day could have been a victim. All those commuters were equal, but that in no way made up for the hideous unfairness that befell some of them.

And while all this was going on, the eyes of the world were on an exclusive hotel several hundred miles north of London, in Gleneagles, Scotland, where leaders of the G8 countries were meeting. As host nation, the UK had declared that the summit would focus on addressing global poverty. I had a great deal of hope for what it might achieve. For the two years leading up to that week, my job at DfID had been helping to write policy for shaping many of the discussions at Gleneagles, including the UK's global strategy for addressing girls' education, which was thought of as the silver bullet of development. By the time of the summit itself, my work over these two years had been edited and negotiated and boiled down to just one line of text: 'we guarantee universal primary education to all children in sub-Saharan Africa'.

I certainly wasn't alone in wanting to tackle world poverty. Thanks to the concerted work of campaigners in the years running up to 2005, the profile of poverty as an existential threat had never been higher. In the week before the summit, over 200,000 anti-poverty activists had arrived in Edinburgh from all over the world to take part in a huge march and form a human chain around the city as a gesture of solidarity with the world's poor.

The scale of the problem can be gauged from the scale of the suggested solutions. To meet its Millennium Development Goal targets for tackling poverty by 2015, the UN had calculated aid should be boosted by $70 billion more in 2006, rising to $90 billion more by 2010. Figures like that could only be raised through coordinated action by the world's richest countries. It might therefore be assumed that the delegates would be left in peace to do their important work.

But poverty is an emotive subject, and anger about it transferred very easily to anger aimed at the summit's guests, many representing countries and institutions whose policies were seen as part of the problem. The summit itself seemed to encapsulate the unfair inequality that angered so many people. According to the Gleneagles booking site, in the month that I wrote this chapter in 2023, just a no-frills week's bed and breakfast would set an ordinary paying guest back over £5,600.

On 8 July, the summit saw what were described by local residents as 'violent and chaotic' scenes. Roads were blocked off by sit-in protests and marches. Police lines were breached by hundreds of protesters, and police in riot gear were flown in by Chinook. A 52-mile fence was erected around the hotel and its grounds to shield delegates from the outside world, and just under 11,000 police officers were on duty or on standby. In total, the police operation to protect the delegates – the largest and most expensive ever seen in the UK – cost £72 million. These were the conditions under which some of the most powerful men (and yes, they were all men) in the world would discuss the crippling problems facing the rest of the global population.

Still, it must be said that not all the hopes pinned on the summit were entirely dashed. Along with other measures, 18 of the poorest African nations had their debt cancelled.

£28 billion was pledged to boost the economies of developing countries. The EU members of the G8 committed to a collective foreign aid target of 0.7 per cent by 2015. And the G8 agreed to the one line of text I had helped write. Afterwards, Oxfam said that the summit did more for development, particularly in Africa, than any other G8 summit before.

But 'more' is a relative term. It is better than 'nothing', but still, in comparison to what the UN had said was necessary, it was calculated the promises of Gleneagles would only halve poverty. In the opinions of campaigners, a lot of the promises fell short. For instance, the commitment to increase aid to 0.7 per cent of gross national income by 2015 was not met by most donor countries. The pledge to provide universal access to HIV/AIDS treatment by 2010 left over 3 million people in low- and middle-income countries without vital anti retroviral therapy. And the promise to cancel debt for the poorest countries did not fully materialise. These unmet targets did little to dampen the seething anger among the summit's critics.

Nearly 20 years later, a summit that could have changed the world forever is mostly remembered for the image of President George W. Bush falling off his bicycle while trying to wave at some police officers.

* * *

Despite seeing some progress in international development throughout my career, I have sometimes felt that the challenge is insurmountable. The greatest factor holding back progress is inequality. Unfair inequality. As described earlier, unfairness and inequality are not the same thing. By inequality we don't just mean the simple fact that some have more, and some less. We are talking about the accumulation of wealth and power in a few hands that will not let go, but instead grasp for evermore. That is unfair; that is what makes people angry.

Income inequality also makes populations unhappy. A powerful few come to exercise a vastly disproportionate influence, usually serving their own interests over those of the broader public. Festering resentment fuels crime and social disorder. Reduced access to basic services such as education and healthcare make people less likely to acquire skills that could help them, which chokes the chances of economic and social mobility and depresses economic growth. Communities already marginalised by race, sexuality or gender take the brunt. A growing sense of vulnerability in the population can produce a fertile breeding ground for discrimination, authoritarianism and extreme nationalism. In 2018, which marked the twelfth consecutive year of decline in global freedom, 71 countries saw a net decline in political and civil liberties.

So inequality clearly needs to be tackled. But as the legacy of many failed egalitarian political movements suggests, creating equality is not straightforward. And, as the experiment with coins showed, sometimes people prefer fair inequality to unfair equality. The first step is to decide how we distinguish the two. Generally, we feel something is unfair when some principle of justice is violated. It is not unfair that someone should suffer an unavoidable accident at work; it is unfair if their employer deliberately skimps on the necessary protective equipment to save costs. It is not necessarily unfair that some jobs are paid more than others; it is unfair if the better off rig the system so that no matter how hard the poorly paid work, they will never prosper.

In their book *American Society: How It Really Works*, social scientists Erik Olin Wright and Joel Rogers say that in order to understand inequality we in fact need to make two judgements: a moral judgement that an inequality is unfair, and then a sociological judgement that this unfairness could be remedied. On top of this is a quadruple assessment: Who deserves what? What kinds of inequalities are justified? What kinds violate principles of justice? And what should be done to redress an injustice? Judgement must be made not just in the name of moral authority, but because when people feel something is unfair, they are more likely to support efforts to change it. Unfairness that makes us angry often arises because someone, somewhere could remedy it, but does not.

So while inequality must be tackled, it must be done so with discretion and discernment. And this is where ideas from physics can help – because, as mentioned at the start of this chapter, if there was no inequality, the universe as we know it would not exist. To understand why, let's meet the 20th century's greatest physicist.

*   *   *

I imagine the schoolboy Albert Einstein sitting in his school classroom in Ulm, in the Kingdom of Württemberg, learning many of the same physics principles that I learned as a boy. Like me, I'm sure he learned about Newton's three laws of motion. And like me, he was probably taught Newton's law of gravitation, which says that every object in the universe attracts every other. But while I accepted Newton's theories without question, for the young Albert Einstein, something didn't quite add up. According to Newton, the force of gravity was felt instantaneously. In other words, if the Sun were to suddenly vanish, all the planets in the solar system would feel the absence of its gravitational pull instantly, and presumably start drifting off into space. But for a young Einstein, this was incompatible with how he saw the universe.

In 1905, he published a paper called 'On the Electrodynamics of Moving Bodies', more commonly known as the theory of special relativity. This paper contained a number of revolutionary ideas, including the famous equation $E = mc^2$, which states that the energy (E) of an object is equal to its mass (m) times the speed of light (c) squared. The paper also postulated that the speed of light in a vacuum was a fundamental constant. In other words, it was always the same. No matter where you are in the universe or how fast you are going, if you were to measure the speed of light, you would always get exactly the same result, which happens to be a staggering 299,792,458 ms^2. In the years since, numerous experiments have been carried out that proved Einstein right.

Crucially, one of the implications of the constancy of the speed of light is that nothing in the universe could ever travel faster than the speed of light. The universe has a hard-wired speed limit. And herein lies Einstein's issues with Newton's theory of gravity: If nothing can travel faster than light, then how could gravity be felt instantaneously? Gravity too has to obey the universe's speed limit. This means that if the Sun were to instantly disappear, we wouldn't experience the lack of gravitational pull immediately. It would take at least eight minutes, which is how long it takes for light to travel from the Sun to Earth.

Faced with this challenge, Einstein embarked on a quest to develop a new theory of gravity – one that would reconcile the principles of relativity with the gravitational force. After a decade of intensive work, Einstein unveiled his next great paper, known as the theory of general relativity. Einstein's goal in writing this paper was to understand the mechanics of how gravity actually worked. What was actually happening to the Earth as it orbited the Sun, or the apple falling from the tree?

To answer this question, Einstein needed a new model. Imagine a large, taut drum skin stretched out flat. If you were to roll a small marble across the drum skin, it would likely travel in a straight line. Now imagine placing a heavy bowling ball in the centre of the drum skin. This ball would likely cause the drum skin to stretch, creating a dip in the surface. What would happen now if you rolled the small marble across the drum skin? Would it still travel in a straight line? No, it would likely follow a curved path. If the marble had enough velocity, it might even travel all the way round the drum skin in some sort of elliptical orbit. In other words, without any direct contact, the presence of the bowling ball changed the path and movement of the marble.

For Einstein, this simple model was an excellent analogy for how gravity works. Instead of a drum skin, imagine the fabric of space and time. Instead of a bowling ball, imagine a star like our Sun. And instead of the marble, imagine the Earth. The presence of the Sun stretches and bends the fabric of space – time and in doing so changes the path of the Earth, forcing it to move in an orbit. The same is true of the falling apple; the Earth stretches the fabric of space-time, which affects how the apple moves. Einstein concluded that matter tells space how to stretch, and space then tells matter how to move. This is how gravity works, and again, in the years since, Einstein's theory has been shown in experiments to be correct.

Einstein's theory of general relativity not only offered a more nuanced understanding of gravity but also paved the way for a new model of the whole cosmos. Before general relativity, physicists used to believe that the universe was a constant size. But Einstein's equations showed that this was impossible; the universe had to be expanding. And if this is true, it means that the further back in time you go, the smaller the universe must have been. And if you go far enough back in time, the universe must have started from an extremely compact, hot state – what we now call the Big Bang.

General relativity provided the framework that led physicists to discover that the universe began some 13.8 billion years ago, as a singular point of infinite density and temperature.

At this moment, all that was and all that would be, existed as a singularity – an infinitely small, infinitely dense, perfectly uniform point. For a brief moment, the universe was completely ordered and everything, everywhere was the same. It was perfectly equal. In a deterministic universe, with nothing to disturb this order, this might have continued indefinitely. But this is not the nature of our universe.

As we've seen, in the quantum realm, there is no such thing as certainty. It's impossible to ever simultaneously know the precise position and movement of a particle; instead they exist as waves of probability. In the everyday macroscopic universe, we don't experience this quantum uncertainty. But in the early universe, this uncertainty had a huge impact. In quantum physics, even what we consider a vacuum is not truly empty. Because of uncertainty, energy fluctuations can spontaneously occur, causing particles to pop in and out of existence. These transient occurrences are known as quantum fluctuations. In the early moments of the universe, these fluctuations led to tiny variations in energy levels. Although these deviations were incredibly small – no bigger than 0.003 per cent – the rapid expansion of the universe at the time stretched and dramatically amplified these fluctuations.

Contrary to popular belief, the Big Bang was not an explosion of matter that expanded to fill an empty space. It was an expansion of existence. The Big Bang created and stretched space itself. As the universe expanded, energy flew in all directions at the speed of light – a million times faster than the explosion of a hydrogen bomb.

The force of gravity in this early universe was so strong that even light couldn't escape. It would take about 300,000 years for the universe to finally cool to a still-balmy 3,000°C, when atoms could finally form the very first elements and light could begin freely travelling outwards across the rapidly expanding universe. Today, that light is still visible to astronomers on Earth and is known as the cosmic microwave background radiation – an echo or 'shock wave' of the Big Bang, as it were.

After 100 million years, the tiny variations caused by the quantum fluctuations led to pockets of energy and matter coalescing. With denser mass in those pockets came stronger gravity. Stronger gravity caused gases to cluster and eventually crush nuclei together, initiating a process of nuclear fusion. This released light, heat and energy. And so were born the first stars.

These early stars were at first roughly evenly distributed. But those stars, like all matter, still tended to attract each other. A few more hundred million years and the stars – not to mention everything else still floating around in the form of interstellar dust and gas – had coalesced together into giant collections that we call galaxies. As the galaxies gathered into themselves, the spaces between them grew even larger.

Fast-forward a few more billion years and the universe was composed of huge, low-density voids, occasionally interspersed with extremely dense regions. Galaxies, clouds of dust and dark matter, the remains of dead stars and the embryonic forms of new ones, sleeting storms of radiation and gravity waves all punctuated tracts of void millions of light years wide. On a cosmic scale, either a lot of something was going on, or almost nothing at all. There were very few in-between bits.

Stars forged heavier and heavier elements in their hearts. And when they eventually died, they exploded as supernovae, flinging their elements back out into the void. Those elements were added to the rich mix of matter floating around in space. Once again, some areas were denser than others, and once again the star remnants were pulled together. More stars were created – including our own Sun – and eventually solid objects called planets were formed.

And on at least one of those planets that we know of – the one you and I live on – conditions were right for the emergence of life.

So it is thanks to those tiny levels of inequality in the early universe – tiny differences in energy levels, in density, in gravity, no more than plus or minus 0.003 per cent – that we have the universe as we know it today. If those initial differences had been much smaller, then there would be no structure to the universe at all – no stars, no planets, just a general homogeneity. There would certainly be no life. And if those differences were much larger, then the universe would have been very short-lived, because the dense regions would have been so dense they would have coalesced into black holes – lumps of matter so dense that not even light can escape from them. The universe would have collapsed back into itself long ago.

That is why many people refer to this as a Goldilocks universe – neither too dense nor too light but just right for structure and life. It is very fortunate for us that the universe began unequal, and continues to be so today. We have gone from a gaseous mass, to a universe-sized cluster of billions of stars, to galaxies, to planets, to one planet in particular and life emerging there.

*   *   *

One particular life form of interest to this chapter is Adrian Bejan. Born in 1948 on the banks of the river Danube in Romania, Bejan is today a professor of mechanical engineering at Duke University in North Carolina, and one of the world's foremost experts on thermodynamics and the same fundamental laws of physics that guided the universe's expansion.

Before studying physics, Bejan grew up under the rule of Nicolae Ceaușescu, one of the most vicious Stalinist dictators thrown up by Communism in eastern Europe after the Second World War. Like all Communist countries, Romania was meant to be a workers' paradise – a living rebuke to the decadent West; an exemplary society where no one lacked anything. Both fair and equal. What that meant, says Bejan, was everyone was equally poor and equally deprived of liberty. No one could say anything that contradicted this official view.

'When I was born, 70 per cent of Romania was peasant and the peasantry was essentially illiterate,' Bejan remembered. 'My father was one of them. If anyone stayed in school in his village, it was up to fourth grade.'

Food was scarce – meat in the shops had all but disappeared. Bejan's father, a veterinarian, decided to raise chickens in the basement, providing eggs to neighbours and friends who had access to few other sources of protein. 'He had a friend, an electrical engineer, design a wardrobe-size incubator for him,' Bejan explained. 'He also had a light box the size of a Kleenex box. There was a bulb inside, and a circular opening against which [from the outside] he would hold an egg. Light would pass through the egg, and he and I could see the growth of the embryo from evening to evening. We were seeing the first blood vasculature forming and spreading on the inside of the eggshell.' The blood vessels were radiating and bifurcating out in all directions from the initial 'node', which was the living embryo. The image was strangely reminiscent of something else that dominated the young Adrian's life.

The family lived near the delta of the Danube, where the river empties into the Black sea through a network of ever-smaller channels. 'Every spring there's a flood after the thaw. I would often go fishing on the riverbank. I discovered that the river had moved. It was not there any more. The river was alive. It was throbbing.'

Bejan eventually drew the parallel, without realising its full significance: The blood vasculature on the eggshell as the embryo grew inside was like the pattern of the Danube delta. 'Absolutely the same design. And, well, that's life, animal and river alike.'

Fast-forward a few decades. Thanks to a kindly uncle, Bejan's father was lucky enough to stay in school beyond the third grade. His uncle took him to a boarding school in the big city and got him a proper education. This later translated into an early boost to his son's own prospects. Bejan also received a good education and at the age of 19, he won a maths contest and a scholarship to the Massachusetts Institute of Technology in the United States. He was awarded a PhD by the same institution in 1975, and has never looked back since.

But, no one ever entirely escapes their childhood and Bejan was no exception. In 1996, he was designing a cooling system for laptop computers. The system imagined by him was in the form of a tree-like structure of highly conductive blades for heat flow, a large channel that branches into ever smaller ones, running all the way through a solid block of metal. The heat of the processor warms the thick blade nearest to it, and the heat begins to spread through the body of the laptop, while cooling the processor.

And once again, he saw a similarity of patterns. The breakthrough realisation was that this form – a large channel that branches off into smaller and smaller ones – occurs in any flow system, whether natural or artificial. To physicists like Bejan, a flow system is any system that involves something moving under its own power from A to B. In nature, it can be the channels of a river delta, or blood vessels on the inside of an eggshell, or branches of trees, or the nervous system of a body. In the man-made world, it can be computer cooling systems, or subway maps or communication networks. The flow of knowledge and data takes on just the same form, Bejan explains, including in 'fields where human perception, or bias and culture, and some other things like religion play a more upfront role'. So, whether in a FTSE 100 company, or a church community, or an army, the flow of information through an organisation will be tree-like, starting with a few individuals at the top, then down through ever-branching levels towards the rank and file.

The branching pattern of these phenomena is not a template they are impelled to follow – rather it is simply the most efficient means for something to flow. When Bejan adopted it to design his cooling system, he says, he was not mimicking nature as such: 'I was predicting nature. I realised that this tendency of the chicken embryo and the river delta was in fact a phenomenon of physics – the tendency to evolve this design in a direction. It was screaming for its own law of physics.'

The result is what Bejan now calls the Constructal Law,[1] and this – just as much as his work on thermodynamics – is why he was awarded the Benjamin Franklin Medal. The Constructal Law says that there is a universal, natural drive for flow systems to evolve towards greater access. It states that for any flow system to survive, its configuration must evolve in a way that allows the easiest access for whatever is moving through it. The alternative is stark. Bejan puts it bluntly: 'When movement stops, life ends.'

There are various conclusions that can be drawn from this.

Systems change and evolve over time in predictable ways, but always to improve flow. A common misconception about evolution is that living creatures began as inferior and are in the process of becoming superior – as though we were on a one-way journey from being amoebas

---

[1] *The Physics of Life: The Evolution of Everything*, Adrian Bejan, St. Martin's Press, 2016.

to becoming superhuman. Because of this, it can seem that evolution has some kind of discernible end point. In fact, evolution merely drives something to become the best fit for its environment. If the environment changes (and it does change) then so must the evolving system. No matter how well adapted it is, any species that rests on its laurels will always, ultimately, fail.

Just as the universe grew after the Big Bang, the process described by the Constructal Law goes in one direction only – from obstructed to easier flow. All flow systems eventually start to show a branching structure that can be described as 'few large, many small'. Big currents lead to ever-smaller streams. The English mathematician John Horton Conway calculated that, across all flow systems, for every big stream, you're likely to see an average of four tributaries emerge.

As Bejan explains: 'The bigger things tend to be the more efficient carriers. Freight is cheaper on bigger trucks. But that's only half the story. The other half is that big things cannot possibly flow over areas smaller than they are. They can't reach into the tiniest pockets. The architecture of any efficient system must comprise many smaller volume elements and branches. The architecture that emerges naturally is a hierarchy.'

Hierarchy is inevitable, and hierarchies are by definition unequal.

The Constructal Law manifests itself in everything and anything that flows, from the fluids inside our bodies to the electricity that feeds our homes. But Bejan's particular interest has been the flow of something else: He tracks the flow of *money*. Bejan's central question: Is it possible to predict the flow of wealth in society? If so, then the questions that follow are sobering. Is inequality inevitable? Do the rich get richer because the laws of physics dictate that they must? And is any of this fair?

Bejan's research reveals that, like every other kind of flow, the spreading of wealth is hierarchical – and, therefore, unequal by definition. A flow system requires large branches feeding into smaller ones. In a lung, the trachea is bigger than the bronchi below it, but not because it is greedy for air. It is precisely as big as it needs to be for the overall size of the flow system. The channels of the Mississippi basin will be bigger and more numerous and complex than those flowing off the Seine or the Thames. So, some people might be rich or even super-rich – but according to Bejan's theories about the physics of flow, that is only a problem for society if the flow of the system is interrupted.

Bejan compares the physics of wealth to a steam locomotive. When stationary, it needs to let off steam because the pressure that continues to grow inside its boiler has no outlet. If there was no valve to let the steam escape –

no way of letting the steam flow out – then the only alternative would be for the boiler to explode. That is the physics
of an economic dictatorship: a rigid flow structure with
no capacity to grow or evolve. There is no hierarchy, no
inequality, but there is also no movement. Communism,
says Bejan, speaking of the system under which he grew
up, meant 'everyone was equal', and it failed. His father
didn't keep chickens as a hobby or for the health benefits
of fresh eggs; he did it because there was no other way
of getting enough food. 'You try to make it one-size-
fits-all, you lose.'

Good news for capitalism? The physics of flow suggests
that the notions of hierarchy should not unduly trouble us.
The inequalities they give rise to are necessary and need
not be unfair.

But 'should not' is a long way from a categorical 'is
not'. As the example above shows, it is perfectly possible
for someone with wealth to misuse it. A critic might say
that Bejan's physics are just as idealistic and impractical
as those of the Communists whose disastrous planned
economy he experienced at first hand. And, as a white-
collar professional living and working in a well-paid job
in the United States, while he might not be at the top of
the hierarchy, he is a long way from the bottom. What
would he say to people who are at the other end – the
people who are living in financial precarity, as so many in

his adopted country are? Or for that matter, how would he speak to a development economist or someone at the World Bank trying to deal with extreme poverty in Africa or south Asia? Are there lessons from the Constructal Law that could provide concrete solutions or paths forward for policymakers?

To start addressing the problem of unfairness in the world, he says, those with wealth and resources must consciously seek out where the need is. They need to become the channels through which wealth flows, for example, towards philanthropy. 'The answer is in plain view. Give a damn. That's the answer.'

* * *

A lot of what has been described here comes close to the free-market dream of trickle-down economics. According to this school of thought, the market should always reign supreme, everything will find its own natural level, and wealth that accumulates among the rich will find its own ways to trickle down to the less rich. The distribution of wealth might remain unequal, but in the long term, everyone benefits. The less government involvement there is in the population's affairs, the better.

As you would expect of a survivor of totalitarianism, Bejan is no fan of unnecessary government intervention.

However, he is no unlimited free-market libertarian either. He is well aware of the good that intervention can do, if appropriately applied. The Constructal Law suggests that while inequality is ever-present in the universe, unfairness does not have to be. Unfortunately, there is nothing in the trickle-down theory to guard against it.

The channels of the Danube delta that inspired the young Bejan were formed naturally, but they could be blocked by fallen trees, rocks or silt. If that happened, they would flood and cause damage. Locals would therefore take great care to see that they remained unobstructed so that the water could flow freely. Occasionally, perhaps a channel had to be blocked for practical reasons. That was fine – as long as another channel was then deliberately created to redirect the flow.

Humans do not have to be at the mercy of the flow. They are capable of blocking and unblocking channels through their own actions. It is only when flow is interrupted, Bejan says, that 'pressure builds and builds until the system dies. The moral: Keep things moving. Increasing flow is what sustains organisms and devices, and stagnation kills them.'

The branching pattern of the Danube delta and the eggshell blood vessels of Bejan's childhood arose out of natural, evolutionary processes – but the cooling system of a laptop that draws on the same physical principles was intentionally designed by a human mind. All follow the

Constructal Law. In short, there is nothing wrong with artificial intervention to ensure a fairer distribution, as long as this too follows the designs dictated by the Constructal Law. Anything else is doomed to failure. So even if you believe in laissez-faire trickle-down economics, you still need to intervene sometimes to keep the channels open.

Reflecting on my time working in international development, our most effective attempts to tackle poverty and rebalance wealth followed the Constructal Law. Aid and philanthropy have been the primary vehicles for addressing inequality for decades, and this was certainly the focus of my work. I wrote strategies for investing billions of pounds of aid in developing countries. But effective aid is about so much more than simply pouring money into a poor country.

Imagine a simple situation – say, everyone in a tiny village of five families is gifted exactly the same amount of money. Bejan's theory shows that even here, you will eventually start to see an unequal distribution of wealth. Individuals are different and so too is what they would do with the same amount of wealth. Luck also plays a role; factors like illness and accidents will befall some but not others. Some families will end up becoming richer, others poorer. As society grows, that non-uniformity grows ever more complex, with greater differences in wealth and associated opportunities.

Bejan shows that wealth injected into any country via aid will ultimately end up being unequally distributed, just like the wealth that was there to start with. Physics always wins. So, if we accept that wealth will always spread unequally, we need solutions that address this proactively and with fairness in mind. We need to direct the flow.

One way to achieve this is through taxes. David Amaglobeli and Celine Thevenot of the International Monetary Fund argue that in advanced economies, direct taxes and transfers of wealth can jointly reduce income inequality by more than one-third. However, just as effective aid is more than simply giving people money, effective taxation is far more complex than just taking it off them. Rather than be used – as it all too easily can be – to plug holes in a leaking financial dyke or to throw money at a problem, taxes and spending programmes need to be progressive and targeted, so that everyone gives proportionately to what they can afford, and the proceeds are directed to the areas where they will serve best.

Tax money could be used to provide universal early childhood education, coupled with increased access to higher education, community colleges, and vocational and apprenticeship programmes – perhaps government-sponsored job training. Tax credits can be provided for research and development. Tax money can also of course pay for social safety nets that stop people hitting rock bottom and give them a necessary boost back to the point where they can start to make their own luck, catching up on the advantages that children of richer parents have had from birth. Studies show that countries that finance more spending on social sectors through a redistributive tax system tend to have more success in reducing inequality. In short, taxes should have purpose and direction.

James Madison, one of the Founding Fathers of the United States, remarked that 'Nothing is above our courage, except only (with shame I speak it) the courage to tax ourselves.' That is the starting point – another way of putting Bejan's 'give a damn'. One specific solution, which was discussed at Davos in 2023, is a progressive wealth tax on global multimillionaires. The very rich usually have a number of income streams in their port-folio, and they have very clever accountants who see that the majority of their wealth is officially derived from the lower-taxed categories. Property wealth is already taxed all over the world but it is usually a flat tax, not progressive – the very rich pay the same percentage on their property as the average citizen. And as different countries have different tax regimes, the property of the super-rich tends to be officially based in areas where the taxes will not be high. So, the recommendation from Davos was to expand property tax for multimillionaires to encompass all forms of wealth, and to make it progressive. Under the model put forward at Davos, their property would be taxed in the same way in every country. Such a tax could generate trillions of dollars each year that could be reinvested in health, infrastructure and perhaps even books in classrooms.

That recommendation, however, was the result of a – perhaps utopian – outlook, one that would require all the countries of the world to act together as one to tackle the problem. Some may think this sounds a bit like some kind of economic totalitarianism, like that Bejan grew up with. In reality, it is far from the case. The failing that Bejan identifies in Communism is that it denied people the freedom to be enterprising with the little money that reached them, whereas strategic and consistent taxation merely makes sure the money does reach them in the first place. After that, it will follow the Constructal Law and flow in the most efficient way.

* * *

Physics has taught me that there will always be inequality. Trying to stamp it out is futile; instead, it is far more productive to tackle unfairness. Inequality of energy powers the universe; without it, stars wouldn't shine, planets wouldn't orbit and we wouldn't exist. This 'inequality' might also be described as differentiation – a necessary condition for the dynamic and vibrant nature of our cosmos.

112

While we should strive to reduce extreme inequality, a truly 'equal' and undifferentiated world is impossible to imagine. Such a world would lack the richness that comes from our differences, both in talents and perspectives. Removing inequality in physics would collapse the universe into a tiny, featureless blob. Similarly, attempting to eliminate all societal inequalities can be oppressive. Instead, focusing on eradicating unfairness allows us to pursue a more just world, whilst celebrating our differences. Inequality is ever-present in the universe, but perhaps unfairness does not have to be.

Chapter 4

THE PHYSICS OF GOING NOWHERE

For centuries, the British government has posted diplomats to its embassies around the world. They are all tasked with fostering relationships between countries and promoting Britain's interests overseas. But not all embassies are considered equal. There's a discreet but well-established system for ranking the attractiveness of overseas postings, and certain cities are considered so unappealing, so difficult and so hard to live in that they are actually called 'hardship posts'. As you would expect, places like Baghdad, Kabul and Kinshasa – cities that have recently been war zones – fall into this category. What you might not expect is that Britain's embassy to the wealthiest, most powerful and most important country in the world was also once technically classified as a hardship post. During my time as a civil servant, I experienced first-hand why.

In 2007, I had recently graduated from business school and had just started a new job in Washington DC. I was working for the World Bank, an organisational melting pot of diplomats and fellow international civil servants. I loved the people and the energy of the city, but when I arrived in midsummer, nothing could have prepared me for the heat – a sweltering and oppressive heat that seemed to drain every drop of energy from my body. Each step on the walk to work felt like a battle against a wall of thick and humid air. Washington used to quite literally be a swamp, which is why until the introduction of air conditioning in the mid-20th century it was indeed considered a hardship – albeit a slightly less dangerous one than a war zone.

Making matters worse, I had made a fatal error before departing the UK. Determined to make a good impression at my new job (and perhaps having read too many John le Carré novels), I had purchased a white linen suit. Slightly oversized, it very quickly became hideously crinkled and adorned with sweat stains. Adding insult to injury, the lightweight fabric clung to my skin, in particular around the seat of my trousers, in ways that were less than flattering to my short and plump physique.

My office was thankfully fully air-conditioned, but the effects of this heat seemed to linger even when I got inside. Sitting in my cubicle, I stared at the blinking cursor on my computer screen for hours on end. Days passed but I made no progress with my work. It was like the city's heat had sapped my vitality. I came into the office every morning, determined that this was going to be the day I would finally make progress. My desk was littered with empty coffee cups and cans of Diet Coke. Yet, by 11:00am most days, the only thing I felt motivated to do was crawl under my desk and take a long nap. I was going nowhere.

Appropriately enough, the report I was trying to write was all about heat. Global warming, to be precise. My time in Washington took place almost exactly a decade after the Kyoto Protocol, or, to use its full name, the Kyoto Protocol of the United Nations Framework Convention on Climate Change – the first internationally binding climate agreement. At the time it was made, it was widely hailed as the most significant environmental treaty ever negotiated, and was certainly the most ambitious. It aimed to reduce across 41 countries the emissions of six of the greenhouse gases that contribute to global warming. My job was to write a report on international progress towards the goals set out in Kyoto.

The Protocol was huge, but not as enormous as the challenge. Unless abated, best estimates suggest that by 2050, a third of the world's population could be living in areas as hot as the hottest parts of the Sahara now. Developed countries, by and large, have the resources to cope with these emergencies, at least for the time being. In the short run, the impact on developing countries will be much harder; in the long run, it will catch up with everyone. It is an existential threat.

Faced with the enormity of this challenge, I struggled to find the energy or the motivation to adequately explain the situation. I didn't know where to start and, as more time passed, felt increasingly panicked by my lack of progress writing my report. And so, I turned once again to the one subject I knew best.

Newton's first law of motion states that an object at rest will stay at rest, and an object in motion will remain in motion with constant velocity unless acted upon by an external force. If this law was applied to people it would tell us that it takes a lot of effort to change human behaviour, especially at the scale of a challenge like global warming. Business as usual is always a stronger force.

The second law says Force equals mass multiplied by acceleration. Just like objects in motion, we too are a product of hidden forces acting on us – mass (the individual) and force (the context in which we operate) leading to acceleration (a particular behaviour). The greater the mass, the more force needed to shift it. There were numerous forces working against the success of the Kyoto Protocol: pressure from business lobbies, the inertia of bureaucrats and politicians, and also the ingrained practices and attitudes of the populace – including you and me.

The third law states that for every action there will be an equal and opposite reaction. Applied to people, this law warns us that every decision will provoke reactions that go against what was intended. In human terms, people just don't like being told what to do, and the more insistently this is done, the more they resist. A large mass moving one way might not be best combated by an equally large mass going the other. Softer forces – like gentle persuasion – may be applied more carefully, and over a longer period, to get better effects.

Thinking about these laws while sitting at my World Bank desk gave me a sense of why the Kyoto Protocol had run into difficulties over the years. Put it down to human nature – which, like the laws of physics, doesn't change. And yet, even if that could be changed – even if believing governments could be sufficiently persuasive, convincing populations and businesses to take the hit for the benefit of tomorrow – I suspected it would not be enough. Just as Newton's laws say that an object at rest will stay at rest, at this moment in time it felt like efforts to tackle climate change, like me, were also going nowhere. Could ideas from physics offer any answers?

* * *

The physics of what we think of as climate change arguably began about 300 million years ago. This was a period of the Earth's history known as the Carboniferous period. During this time, the Earth's climate was warm and humid. The planet was home to vast forests and swamps. When the plants and animals that lived in them died, their remains were buried under layers of sediment. Over time, the heat and pressure from the overlying sediment caused the organic matter to transform into what we now call fossil fuels. Coal formed from the remains of ancient ferns, trees and mosses. Oil formed from the remains of tiny marine organisms, such as plankton and algae. And natural gas, the lightest and most gaseous of the fossil fuels, formed from the same organic matter as oil, but was subjected to more heat and pressure.

For millions of years, these fossil fuels sat dormant under the ground. We knew they were there and we used what we could get hold of, but we barely scratched the surface of the deep reserves. But 400 years ago, a series of events changed everything.

In the 16th and 17th centuries, the British navy was facing a growing threat of invasion from Spain. Faced with this danger, the diplomats and military leaders of the time ordered an urgent expansion of Britain's fleet. This increased demand for ships led to an increased demand for timber, which eventually led to deforestation. This deforestation led to a shortage of charcoal, which had at the time been one of the most common fuels. An alternative energy source was needed, and so began an increase in coal mining. Deeper and deeper shafts were built, each with a greater chance of flooding. In the late 17th century, Thomas Savery and later Thomas Newcomen invented the first practical steam engines, which could be used to pump water out of these flooded mines.

In the 18th century, as the price of coal dropped, demand rose and new coal-powered applications such as lighting for cities emerged. Meanwhile, the growing demand for steam engines triggered important design innovations that sharply increased their efficiency, reduced their cost and broadened the range of their applications. Metallurgical innovations driven by this resulted in the invention of Bessemer steel, which allowed the construction of cheap steel railway networks and their associated infrastructure – not to mention modern iron battleships.

When coal is heated at high temperatures in the absence of oxygen, a new high-carbon fuel called coke is produced. This process, known as pyrolysis, also created a waste product called coke oven gas, which at first was just flared off and discarded. Inventive minds soon realised, though, that this waste could be used as a fuel itself.

In the 19th century, a new kind of engine was invented, which could burn fuel inside cylinders and harness this hot, expanding gas to drive the pistons. This was the internal combustion engine. At the same time, the whale oil used in household lamps was replaced with kerosene, which was made from crude petroleum. The distilleries that made this kerosene realised that their waste product could also be put to use. This was gasoline, which turned out to be the most valuable commodity yet, perfect for use in those new internal combustion engines.

And so was born the Industrial Revolution.

Fast-forward to today and these same fuels power many of the activities at the heart of popular conceptions of a 'good life' – flying, driving, eating meat, living in larger and warmer homes – and the infrastructure necessary to support them. They are incredibly popular and, as a physicist, it's easy to see why. Fossil fuels are phenomenally effective at generating energy – weight for weight, burning petrol delivers 1,000 times the energy of torch batteries. That's why the exploitation of fossil fuels has been a primary driver of economic growth for the past 250 years.

Unfortunately, it's this act of burning fossil fuels that produces greenhouse gases. There are many different greenhouse gases, including water vapour and methane (a by-product of the production of coal, natural gas and oil, and also a result of organic decay), but the main offender is carbon dioxide, or CO_2, which is created whenever carbon is burned in oxygen. A one-gigawatt coal plant burns a ton of coal every 10 seconds, producing three tonne of CO_2 – it grows heavier because of the addition of oxygen to the mix. World total power production is about 1,000 gigawatts.

You don't have to step into an actual greenhouse to feel the greenhouse effect. To experience the sharp difference in warmth and humidity, just get into a parked car on a warm day. The principles are the same. The greenhouse effect happens when more energy gets in than can get out, which is why the solution with a parked car is to open the windows. Hot air is at a greater pressure than cooler air, so it rapidly dissipates outside. This process is called convection. Earth is not surrounded by windows, but gravity has pretty much the same effect. It pulls the air down, keeping it close to the surface. Unlike a car, there's no way to open this window. Warm air can't just rush out into space.

This doesn't mean, though, that heat can't leave the atmosphere. The Earth's surface is constantly absorbing energy from the Sun in the form of visible light. This energy heats up the Earth's surface and then bounces back towards space in the form of infrared radiation, a type of electromagnetic radiation with wavelengths longer than visible light but shorter than microwaves. The Earth's atmosphere, which is 99 per cent nitrogen and oxygen, doesn't absorb infrared heat, and so in theory this radiation could just escape into space, plunging the average temperature on Earth to minus 11 degrees, freezing the surface of the ocean and making the world pretty uninhabitable. Thankfully, there are two things that stop all this heat escaping. One is cloud cover, which reflects infrared radiation back towards the ground. The other is greenhouse gases.

Greenhouse gases are only present in the atmosphere in trace amounts. From AD 800 to the late 1800s, they were consistently just 0.028 per cent of our atmosphere – or 280 parts per million – which shows just how effective they are at keeping in heat. And here lies the problem. In the last century, CO_2 levels have increased by 100 parts per million. This may not sound that much in absolute terms, but even a small increase is effective in trapping more infrared radiation in the atmosphere and driving global warming.

Considering the long view of history, an increase over just the last hundred years may sound insignificant. It's true that the world has heated up and cooled down before. But just one century at an extra four degrees could irreversibly set back our civilisation, upending our societies and economies.

Our ability to make predictions about climate change has been greatly enhanced by the development of climate modelling as a science. These models rely on our understanding of heat and energy. And that understanding is based on physics, specifically the laws of thermodynamics.

* * *

Newton's laws of motion provided a comprehensive framework for understanding the behaviour of everyday objects in the real world. They represented a deterministic world-view, whereby if the initial conditions of a system are known with sufficient precision, then its future evolution can be predicted with certainty. However, as scientific enquiry progressed, cracks began to appear in this deterministic approach.

While Newtonian mechanics provides a powerful framework for understanding the motion of individual macroscopic objects – in other words, things you can reach out and touch, or at least see – it became clear in the 19th century that it was far less capable of describing the behaviour of ephemeral phenomena like heat and energy. That's because they involve the movement of, not just one or a few, but millions of particles at the microscopic level. A new approach was needed, one that was less about precision and more about probabilities. Physicists were beginning to realise that sometimes it was less important to understand what any individual microscopic particle was doing, and far more helpful to know the average behaviour of the system as a whole.

It was this insight that led to the birth of the branch of physics known as thermodynamics. Pioneering work by scientists such as Sadi Carnot, Rudolf Clausius and James Clerk Maxwell embraced a more probabilistic and dynamic world-view. Their work led to the development of the fundamental laws of thermodynamics, rules that were capable of describing the behaviour of heat and energy in terms of statistical averages and probabilities.

The first of these laws is quite intuitive: Energy cannot be created or destroyed. This means that for a closed system – in other words, a system where energy can't get in or out – the total energy at the end of any process has to be the same as the total energy at the start. To illustrate this, consider what happens when you are driving your car. As the car is moving, it has a certain amount of kinetic energy due to its motion. As you approach a red traffic light, this energy must be lost in order for the car to come to a stop. When you press the brake pedal, the pads press against the wheels, and this friction slows down the car. It also creates heat. And at the end of the process, taking the most simplified view, the energy of this heat will be the same as the car's initial kinetic energy. That is the first law.

We can also consider another everyday example. If you leave a cup of hot water on its own for a few minutes, it cools down – or appears to. What actually happens is the heat from the water goes into the air of the room, so that the water gets cooler and the room gets very slightly warmer – but as the heat of the water in one small cup is dissipating throughout the volume of a much larger room, you can barely tell. Eventually, both room and water will be at the same temperature, which will be the combined temperature of the room plus hot water to start with.

Why does it work that way, though? Why doesn't the water suck warmth out of the cooler air and get even hotter? As long as the total energy in the room stayed the same, this seemingly would not contradict the First Law of Thermodynamics.

This is why there's also a Second Law of Thermodynamics. In layman's terms, the second law says that heat goes from the hotter to the cooler, and only in that direction. Scientifically, it is called entropy; colloquially, 'disorder increases'. We started with order: a cool room and a cup of hot water, energy neatly contained in one place. We end with a more disordered state because heat is dispersed throughout the room. Result: a net increase in entropy overall. Entropy is irreversible and it always moves in one direction. Entropy yesterday was always less than entropy today, and that is always less than it will be tomorrow; this directionality is often referred to as the Arrow of Time.

For centuries these laws of thermodynamics have beautifully explained the behaviour of heat and energy. They are also the foundation of the physics of climate change. Understanding these laws allows us to model the Earth's energy balance, predict temperature changes, and understand the mechanisms behind greenhouse gas effects.

Yet despite our understanding of climate change, by the time I was writing my report, many Western countries had already fallen significantly behind their Kyoto targets. In 2007, the European Union had only reduced its greenhouse gas emissions by around 2 per cent from the 1990 baseline compared to the 8 per cent reduction target set for the 2008–12 commitment period. The United States, which had signed but not ratified the Protocol, saw its emissions increase by approximately 17 per cent from 1990 levels by 2007, despite having a target of a 7 per cent reduction. Today, we are still grappling with meeting those targets, with progress varying significantly between countries. While some nations have made substantial strides in reducing emissions, others continue to struggle.

Ultimately, the Kyoto targets required more sacrifice than Western populations were prepared to take on – no one enjoys wearing a hair shirt, however good the cause. Climate change is not just about science. It's also about economics. And that's perhaps why, in recent years, some physicists have been turning their attention to economics.

* * *

The late Robert Ayres was one such physicist. I corresponded with Ayres when I started writing this book, but he sadly passed away before I had finished. He had been emeritus professor at the non-profit business school INSEAD in France and his work had seen him nominated for a Nobel Prize in Economic Sciences. He wrote extensively on the parallels between thermodynamics and economics, arguing that both disciplines are inextricably linked. Like any good physicist, he had taken a first-principles approach.

Consider an oilfield or a coal mine. Wealth does not arise from the simple existence of these resources but rather from their transformation into goods and services. It is no good just to own an oilfield; wealth arises when the oil is pumped out. A mine is no good to anyone unless the ores in it are dug out and smelted. Or as Ayres puts it, 'Wealth in human society is the result of conscious and deliberate reformulation and dissipation of energy and materials.'

So far, so conventional. Any mine or oilfield owner would probably agree. Resources that are not being exploited only have potential value. But where Ayres and mainstream economics part company is how this value is created.

Economists traditionally focus on two things – capital and labour – as the primary drivers of production. In other words, getting stuff done – whether building bridges or selling cars – is all about how much money and how many people you have. But Ayres argues that these same models ignore one crucial thing: energy.

To understand why this matters, we must again take the long view. Our ancestors learned how to use energy from the environment thousands of years ago. The first key discovery may have been learning how to make and tame fire – creating it from the friction between sticks, or making sparks by banging rocks together, rather than waiting for lightning to strike something dry and flammable. Fire not only made food more digestible but it also drove the very first industries – firing clay pots and smelting metals.

Over the centuries, humans continued to exploit the apparently limitless energy available – the Sun, wind and running water. Comparing the almost negligible historic human demand for these resources with what must have seemed like an infinite global supply of energy, it is easy to see where the classical economic view arose. Of course there will always be more! And for most of human history that has indeed been the case.

Today this sense of 'energy optimism' is at the heart of modern economics. Economists assume that the amount of energy available to make a bridge or a car is largely unlimited, and therefore they don't include it in economic models. It is an 'intermediate good', which is created by capital and labour. Once the end-point goods exist, the energy it took to make them is quietly written out of the equation.

But this is not Ayres's view – or, he would say, the view of the laws of thermodynamics. We have learned that the first law states that energy can never be created or destroyed, but only transformed or transferred from one form to another. It must therefore be present at the start of the process, and it does not just conveniently vanish at the end of it. The second law states that entropy – a measure of the disorder in a system – increases. In practical terms, this means that over time, energy will be dissipated and there will be less and less available for doing useful work.

Entropy in the production of goods usually shows itself in the form of waste – waste heat, waste by-products – which is simply not factored into most economic equations. According to Ayres,[1] however, the laws of thermodynamics mean that waste and its consequences are very real. Taken together, these laws have implications for the use of resources.

To help understand this further, Ayres states that 'energy' is actually the sum of two components that he calls 'exergy' and 'anergy'. The formal definition of exergy is 'the maximum amount of work that can be extracted

---

[1] 'How Economists Have Misjudged Global Warming', Robert U. Ayres. Reprinted from *World Watch Magazine*, September/ October 2001. https://cooperative-individualism.org/ayers-robert_ how-economists-have-misjudged-global-warming-2001.htm.

from a material by reversible processes as it approaches thermodynamic equilibrium with its surroundings'. Put more simply, exergy is useful energy. It's the energy that can be put to productive use. In fact, when economists or engineers talk about energy they usually mean exergy – the useful bit. Likewise, when we talk about the availability of resources, we are actually measuring exergy – the amount we can use.

Similarly, when physicists talk about work, they are not referring to labour, but to the exergy that is put into something to achieve an effect. It comes in many forms. A piston compressing gas, or a wheel operating a winch uses exergy. Exergy is needed for the processes acting on a piece of iron to produce chemical changes so that you end up with a piece of steel. Adding heat to fluid so that it evaporates, driving a steam engine, requires exergy.

Machines are powered by the exergy of their energy source. Humans and animals are powered by the exergy of biomass, also known as food. Very often an animal will be used to power a machine too; historically this was one of the main sources of machine energy, which is why we still use horsepower as a unit of power measurement. The exergy of animal-operated machines was first exceeded by inanimate machines in the US in 1870, but it was only during the 20th century that the exergy of fossil fuels overtook biomass exergy.

Meanwhile, the second law tells us that there will always be some energy lost into disorder, and that brings us to our other type of energy. Anergy is non-useful energy. It's the energy that's wasted. There will always be some anergy – that can't be helped – but the more efficient the process, the less there will be. Consider, for example, an industrial process that creates a lot of waste heat. The power source driving the process is exergy. Waste heat that just goes straight up the chimney is anergy.

You cannot have exergy without anergy. But how we manage the balance between these two different forms of energy is critical. Ayres argues that applying the principles of thermodynamics to economic systems can reveal new opportunities for improving the balance between the two. The novelty of his argument lies in the concurrent application of these principles to both fields.

Consider again an industrial process. Intuition tells us that if some of the heat escaping up the chimney can go on to drive further processes – turn a turbine or just warm a building – then it can be transformed from anergy to exergy.

In fact, heat is a very common by-product of most mechanical, chemical and electrical processes, and it is rarely reused as efficiently as it could be. Ayres says that the average efficiency of electric power generation from heat in the US reached about 32 per cent in the early 1960s – and has stayed pretty much at that level ever since. One reason is that power plants tend to be remote, making it logistically challenging to capture and utilise waste heat. Transporting this heat to where it can be used is often impractical. But that doesn't mean it can't be done.

While some processes produce too much heat anergy, others fail to turn naturally occurring heat into exergy. There are freshwater-producing desalination plants being constructed all around the world – California, the Gulf states, Saudi Arabia – that need heat to power their processes. Solar heat is not only perfectly adequate for this purpose, it also tends to be plentiful in precisely those areas where desalination is needed. Yet so far, very few people are putting two and two together. Most plants generate the requisite heat from burning fossil fuels.

Ayres starts with the fact that the products we consume were produced by a process that also created waste. Many of those products will also become waste at the end of their useful life. The more an industrial society produces, the more waste it generates. Ayres states that as our economy becomes more complex, so do the processes involved behind the scenes, and the default tendency for these is towards inefficiency, because the more complex those processes, the greater the waste – and the greater the exergy that could be put to better use.

* * *

A few years before my time in Washington, President George W. Bush had commissioned a national energy policy for the United States. Known colloquially as the Cheney Report, named after Vice President Dick Cheney, this policy argued that the only way to keep the US economy healthy was to greatly increase its supply and consumption of coal, oil and natural gas. For Ayres, this report epitomised economists' failure to consider the laws of thermodynamics. He identifies three mistaken assumptions.

First, the report assumed that investment choices are always optimal, because firms will automatically maximise profits and consumers will automatically maximise utility. There is no need for energy-saving, because no energy is wasted. There are many ways in which these assumptions are shown to be faulty. The sorry tale of the subprime mortgages scandal that led to a worldwide recession earlier this century is just one indication that investments are by no means always optimal.

The report's second assumption was that energy is never scarce; increased consumption as a result of economic growth will be met by endless supply, with no increase in price. Simple mathematics shows you the flaw here. If you plant one new tree for every two that you cut down, you will eventually run out of trees. If resources aren't being replaced at the same speed they are consumed, they will eventually become depleted. The oil reserves we voraciously consume took millions of years to develop and can only be replaced by natural processes taking millions more.

Finally, because we are generally considered to be economically better off than our ancestors, the report assumed constant, steady equilibrium growth – around 2.5 or 3 per cent per annum. It is this assumption that drew Ayres's greatest ire. For him, the economy is never in equilibrium because equilibrium is thermodynamically impossible; disorder always increases. The economy is subject to entropy. And growth is never smooth and steady. Long-term growth, as well as progress towards sustainability, requires radical and disruptive innovations, resulting in new products and services.

Since industrialisation, economic growth has always been predicated on innovations requiring cheaper and cheaper energy. But Ayres stresses that the value we attribute to energy does not change in a regulated, linear way. The first person to discover that the rocky bit of ground they owned exuded a strange black substance was not guaranteed riches. Its value has gone up and down alongside the development of industry. In Ayres's view, energy powers the economy, but economic activity also arises from the availability of energy. It's a two-way process.

'Increased demand for cars stimulated demand for paving machines, traffic lights and personal-injury lawyers – and eventually for Gulf War weapons, traffic reporters, drive-in fast-food restaurants, and advertising copywriters,' says Ayres. Economic growth is a positive feedback cycle. Availability of goods and services triggers demand; demand triggers investment and more production; economies of scale lead to savings; savings lead to price reductions; declining costs stimulate demand and lead to the creation of whole new goods and services, whereupon the loop begins again.

Economic history and technical progress have been largely driven by this feedback cycle, and the changes can be abrupt and disruptive. Investments do not always pay off: Fortunes were poured into the development of England's canal network in the 18th century so that industrial goods could be transported easily by water, but within a generation the network was falling into disuse as goods shifted over to the new rail network.

Economists have not consistently been able to predict many of the great changes: steam power to electric, narrowboats to steam engines, ocean liners to airliners. Change creates new assets while also wiping out old ones, meaning that economic growth might show a net increase, but there are a lot of ups and downs, and both losers and winners, along the way.

Innovations do tend to lead to greater efficiency as new processes and technologies are devised, but greater efficiency alone is not all that drives growth. A car with power steering, air conditioning and electronically controlled timing may be far more innovative than an equivalent model from 50 years ago, but the processes by which the extra features were produced – which depend on technologies that did not even exist 50 years ago – may be deeply wasteful. The most beneficial approaches are those which identify the most productive use of exergy.

For instance, consider whether it would be more energy-efficient to keep driving your old car or to buy a brand-new electric one. Studies have shown that the production of electric vehicles (EVs) is energy-intensive and can result in significant carbon emissions. A 2018 study by the IVL Swedish Environmental Research Institute found that producing the battery for an EV results in 61 to 106 kilograms of CO_2 emissions per kilowatt-hour of battery capacity. This means that a 100 kWh battery could result in over 6 metric tonnes of CO_2 emissions just from its production, equivalent to driving a gasoline car for just over a year.

There is no doubt that many industrial processes have become more efficient over the years – the energy crisis of the 1970s came as a wake-up call all around. It is ultimately the definition of economy to try to do more with less. But thermodynamic efficiency gains still tend to be concentrated at the start of the process – whether burning fossil fuels or generating energy from renewable electric power – while still allowing waste heat to be lost at the end just as before. Innovation must result in major reductions in raw materials and energy consumption, and improvements in the way that waste is used and treated to make it more useful.

*　*　*

Looking back at my time in Washington, Ayres's theories suggest that the Kyoto Protocol was perhaps destined to fail. Kyoto was built on mainstream economic theories, with the assumption that energy is an intermediate good and not a vital and tangible component of the entire economy. Thermodynamics teaches us that energy is something fundamental; something that is neither created nor destroyed by human activity, but only transformed. We can replace one less efficient technology with another, more efficient one – again, for example, transition from oil to electricity – but this will not necessarily lead to an overall gain if the balance between exergy and anergy is still mishandled.

Ayres's perspective effectively rewrites today's economic playbook. Future economic models need to be built on not just two, but three equally important factors of production: capital, labour and *energy*. And they must recognise and account for the harmful by-products of consumption and production, such as wastes and emissions.

It is not just classical economists who are getting it wrong, Ayres says. Counter-intuitively, the sustainability movement could also take a page out of thermodynamics. Over recent years, we have seen environmentalists advocate for costly energy sources and materials to be replaced with more environmentally friendly ones. But if we are just shuffling resources around, these changes will also not work. Environmentally friendly energy sources are subject to the same thermodynamic laws as the costly ones; they are not infinite and they also lead to waste. If these are not handled properly then the same mistakes will be made. It is only by putting energy at the heart of policy – and any future treaties on global warming – that world leaders can hope to address the root economic causes of unsustainable growth.

Traditionally, the hallmark of a sound economy is a healthy circulation of money driven by commerce. But how we make our money is ultimately more important than how we spend it. If our goods and services leave a vast trail of waste then, regardless of the economy's apparent health, the world is a poorer place. All our problems ultimately come down to our inefficient use of exergy.

Many of the greatest changes to our industries and economies were driven by crisis. Innovation only occurred when the old way of doing things no longer worked or was insufficient. During the Second World War, the urgent need for advanced technology led to significant developments in radar, jet engines and nuclear energy. The oil crises of the 1970s spurred advancements in fuel efficiency and the exploration of alternative energy sources. More recently, concerns about climate change and resource depletion have accelerated the development of renewable energy technologies such as wind and solar power. Crises can catalyse rapid innovation and drive the adoption of new technologies that reshape our economies and societies.

But the kinds of insights that physics and the other sciences offer mean that we now have more foresight than our ancestors. Now, when we see crises coming, we can make changes ahead of time. There is no question that we could be investing in more efficient equipment designed to maximise the use of exergy – machines running on recycled energy, better insulation and so on.

But these changes cost. Businesspeople, when they see that word, tend to react negatively, either stopping or slimming down the expenditure, which reduces its vital impact. Few are willing to invest in anything that takes more than two or three years to pay off.

To make the cure acceptable, therefore, it must be shown that it is far better than the disease. As Ayres points out, costs might be unpleasant to look at on the balance sheet now, but future returns can be immense as scale economies kick in – not to mention the other long-term benefits like new products and services leading to even newer applications. Economies emerge far stronger than before. This has happened before – for example, with the invention of electric lighting, telephones, cars, aircraft, television, electronics – and it can happen again. The difference is that many of the advantages of those new technologies were unforeseen. It will help if now they can be planned for and taken into account.

As it is, though, the costs argument tends to get hung up on who will pay for it now and how widely the benefits will be shared later. These are highly political questions, and the more political something becomes, the more likely it is to be kicked into the long grass for a later generation. As sure as Newton's laws of motion, human nature will not change, so the environment in which humans operate must be changed instead. What is needed, therefore, is language and an understanding of the situation that will satisfy sceptical investors who would frankly prefer to stick with business as usual.

If we continue along our current path of extremely inefficient energy use, we are headed for social, economic and environmental catastrophe. Sustainability is a topic generally associated with more liberal and left-wing polit-ical regimes, but I believe Ayres's outlook makes sense all the way along the spectrum from left to right. Almost everyone recognises the ultimate need to balance the books. It's a question of making sure that all the relevant factors are recorded in those books. Integrating the laws of physics with economic theory allows us to do that.

As for me, thermodynamics taught me a lot about my own energy use. As I slogged away at writing my report, I realised it didn't matter how much coffee I drank; trying to simply boost my energy levels wasn't enough. Instead, I needed to increase my ratio of exergy to anergy. Exergy represents the maximum potential work that can be extracted from a system. For me, this was equivalent to the peak motivation and focused energy I possessed first thing in the morning. On the other hand, anergy signifies wasted energy, akin to the depletion of motivation and mental fatigue experienced as my day progressed.

Over the years, I have found tactics – like waking up earlier, getting more exercise and hiding away distracting devices – that have helped me increase and make the most of my exergy. I have also come to realise that, at least within the human metabolism, exergy comes in cycles. There comes a point when any further effort becomes anergy – wasted energy. The key is to recognise these points, and not resist them. Sometimes it's okay to take a break or stop work altogether. Along with physicists and economists, we all need to learn how to manage our energy.

Chapter 5

THE PHYSICS OF BREAKING UP

Before meeting my wife, I had a relatively short – but volatile – history of dating. I had one or two good relationships where there was some give and take and enough trust to survive the occasional setback, and where actions were seen in the best light. Unfortunately, I also had a few that were less positive; relationships that often felt one misplaced word away from a blazing row, where neither one of us was ever quite sure what the other was thinking, and that more often than not ended with me getting dumped. There are a lot of ways to look at relationships, but as a physicist, I think the difference between good and bad relationships comes down to one thing: stability.

The good relationships were inherently stable. A physicist might compare the two people in a stable relationship to two balls in a valley. Imagine what would happen if you gave either of these balls a small nudge. What would happen? They likely wouldn't go very far. The force of gravity and the curvature of the valley would mean that they would roll back down, meeting again in the centre. Two balls in a valley is an example of what physicists call a stable system.

The bad relationships on the other hand were characterised by a lack of stability. Imagine the same two balls but this time balanced on top of a hill. What would happen now if you gave either of them a nudge? They are the same balls, experiencing the same forces, but this time they would behave very differently. They would likely move a lot further, maybe careering all the way down the hill. This is an example of an unstable system.

You don't need a history of failed relationships to intuitively grasp this notion of instability. Whether it is emotions, financial markets or nuclear reactors, instability can be defined in a similar way. It can be understood as the tendency for individuals or things to behave in an unpredictable, volatile, changeable or erratic manner. Like the ball on a hill, when things are unstable, the smallest nudges can sometimes lead to big changes.

Instability has been evident in a very public way in recent years in one specific – but utterly fundamental – aspect of our society: our democratic elections. On 23 June 2016, I, like so many others, went to bed with a feeling of certainty. It was the day of Britain's historic referendum to leave the European Union, and all signs pointed to a decisive Remain victory. The day before, an online Populus poll – the last conducted before voting began – gave Remain a 10-point lead, 55 to 45. The last thing I remember reading before shutting my laptop on that Wednesday night was an election forecasting expert saying that 'if Leave won, it would represent the biggest polling error in 25 years'.

Membership of the European Union had long been a thorn in the side of British prime ministers. The UK hosted its first presidency of the European Union in 1977, the year I was born. At the time, there were just nine EU member states. But even then the relationship was far from easy. Britain was seen as the 'sick man' of Europe, with a poor economy, over-mighty trade unions and, in James Callaghan, an elderly prime minister on the verge of losing power. Of course, things weren't much easier for his successor. When I was 13 years old, my school headmaster announced on the tannoy system – while, I strongly suspect, holding back his tears – that Margaret Thatcher had resigned. She had been brought down, in part, by those in her party who disagreed with her sceptical approach to Europe.

In the years since, the disagreements in the Conservative Party about EU membership grew louder and louder. And in 2013, Prime Minister David Cameron finally declared that the matter would be settled once and for all. He promised that, if his party won the next general election, they would hold a national referendum on whether to remain in the EU. What's more, although the referendum was legally non-binding, he promised that his government would implement the result. Sure enough, the Conservatives won, and in February 2016, formal campaigning for the Brexit referendum officially began.

The four-month campaign was deeply polarising. The Leave side argued against all the things that it believed had gone wrong over four decades of EU membership. After Brexit, British companies could, they claimed, make trade deals around the world on their own terms. Tighter immigration controls could be imposed. The sovereignty of Britain's courts and Parliament would finally be restored. And, of course, leaving the EU would free up vast sums of money. The now infamous 'Brexit bus' claimed that leaving the EU would save £350 million a week, which could instead be invested in the National Health Service.

Against this, countless economists, academics and business leaders argued that leaving the EU would be nothing short of a disaster for the UK. The Remain side argued that, after a generation of membership, the UK's economy and destiny were too closely tied to the EU for the country to be able to pull out without irreparable damage. Many Leave claims – for example, the compromises on British sovereignty, or the money that would be freed up – were simply refuted as being barefaced lies.

On polling day, the ballot paper asked a simple question: 'Should the United Kingdom remain a member of the European Union or leave the European Union?' Voters were given the option of two unambiguous answers: 'Remain' or 'Leave'. It was the most important decision the UK had made about its future in my lifetime.

Despite the undoubted zeal of the Leave campaign, I had many good reasons to feel confident that Remain would win. The referendum was seen as a sop to the right wing of the Conservative Party, promised by Cameron to shut them up even though he clearly didn't believe he would lose. The Conservative government was officially pro-EU, and no referendum result in the UK had ever before gone against the policies of the ruling government. The Remain campaign was a coherent movement across the political spectrum with one clear promise – business as usual – while the Leave campaign was a hotchpotch of different causes, all promising different things, so no one really knew what the implications of leaving would be. The British public's sentiment towards Europe was generally seen as being favourable. Membership of the EU was all that anyone my age had ever known.

Yet on the morning of 24 June 2016, I and the rest of the country awoke to discover that Leave had won. 51.89

per cent had voted to leave, 48.11 per cent to remain. It was indeed the biggest polling error in 25 years. It was also the manifestation of an unseen but deep instability.

It is not just romantic relationships that can be unstable. Relations between countries, and within countries between leaders and the led, can be equally volatile. Increasingly, the collective decisions we make as a society at the polls have become harder and harder to predict, swinging wildly and unexpectedly from one extreme to another. Can physics make sense of what's going on?

*　*　*

Arguably no one played a more influential and divisive role in the victory of the Leave movement than the campaign's chief architect, Dominic Cummings. He has written much in his blog about how the Leave campaign won. The use of data was critical. The campaign integrated information from social media, online advertising, websites, apps, canvassing, direct mail, polls, online fundraising and activist feedback in ways never done before. But for Cummings, the real secret of success was in hiring the right people. They did not come from traditional fields like politics, business or economics. They were physicists.

The Leave campaign was unique in that it put almost all its money into the hands of people who were normally more likely to be found analysing quantum particles, not quantitative polls. The experts in physics were recruited to do data analysis. Physicists are, of course, numerate, so it's easy to see how their skills would be transferable. But for Cummings, the value of physics extended beyond number crunching. It was about a way of thinking. Writing in his blog, Cummings quoted the late great physicist Richard Feynman's advice: 'The most important thing is not to fool yourself, and you are the easiest person to fool.'

For the Leave campaign, Cummings says physicists were a 'hard floor on fooling yourself' and he empowered them to challenge everybody, including himself. Despite having next to no experience in politics, the physicists working on the campaign had an enormous influence on how decisions were made. They formed their views, according to Cummings, by applying to politics the same set of skills they would to any physics problem: identifying the issue, breaking things down to their simplest form, coming up with and testing a hypothesis, questioning results, and then forming a conclusion. In other words, they applied the scientific method. Cummings felt so strongly about the value of physics that he wrote, 'If you are young, smart and interested in politics . . . study physics. You can always read history books later, but you won't always be able to learn physics.'

Despite Cummings's advice, these days you don't find many actual physicists in politics, either as analysts or as politicians. Armen Sarkissian is one of the few. Sarkissian was president of Armenia from 2018 to 2022 but, before entering politics, he was a theoretical physicist, a colleague of the late Stephen Hawking at Cambridge and winner of the Lenin Prize for Science. He believes that politics should be viewed in terms of the laws of quantum physics, and even coined a new phrase to describe this: quantum politics.

In Sarkissian's view, our interpretation of how politics works needs to be updated to reflect the way that our own understanding of physics has changed over the years. As we have learned, Newton's classical view of the world was linear, predictable, even deterministic. By contrast, the quantum world is highly uncertain and interconnected. The uncertainty principle says that it's impossible, by definition, to measure both the position and momentum of a particle, such as a photon or electron, with perfect accuracy. The more we know about the particle's position, the less we know about its momentum and vice versa. In the quantum world, the very act of observation changes our reality.

Sarkissian compares this phenomenon to the way in which today we all receive individualised news feeds on social media platforms, each serving up different interpretations of events. The act of pausing to read a news item will cause the algorithm to serve you more content on a similar theme.

These personalised feeds create echo chambers where the information we consume can reinforce our existing beliefs, further polarising opinions. Just as in quantum mechanics, where the act of observing a particle changes its state, in politics, the act of consuming information will change the information we go on to receive, altering our perceptions and behaviours evermore. I don't know whether Cummings and his team understood quantum physics as well as Sarkissian. But their use of data and highly targeted social media throughout the Brexit campaign suggests that they intuitively grasped its implications.

This quantum view of politics also perhaps explains why polling is inherently unreliable. The act of measurement – in this case, asking people how they plan to vote – changes the outcome. Imagine a group of parents in a focus group. Few will admit to buying junk food. Yet McDonald's is not going out of business anytime soon. In surveys of what people search for online, few claim to regularly look for pornography. Yet it's one of the most common online search queries. What accounts for these discrepancies? People often misreport their behaviours to present themselves in a favourable light, a phenomenon known as social desirability bias. Traditional polling can be flawed because people are not always truthful. The lesson from physics: Measure what people actually do, rather than what they say they do.

Armenia itself is a nation of 3 million people within its borders but, crucially, it has an expat population of some 10 million – native Armenians who have moved abroad, or their descendants who still regard themselves as Armenian. In 2018 the expat community mobilised opinion against prime minister and former president Serzh Sargsyan so powerfully through social media that he was forced to resign. Sarkissian describes an event like this – having a head of state driven from office by a population not even resident in their country – as having a 'quantum quality'. It is an outcome that could never have been predicted within the framework of traditional politics.

Sarkissian's quantum view of politics draws powerful analogies. In recent years, a new generation of physicists have gone even further to apply theories from physics to understand the decisions we all make during elections in more detail.

A few months after the Brexit referendum, Alex Siegenfeld was a first-year physics doctoral student at MIT. He was 23 years old and the initial focus of his PhD was on the physics of batteries and solar cells – technologies that could one day solve important problems in the energy sector. Siegenfeld had always had a passing interest in politics and economics but had never considered it as a full-time career. In 2016, however, that all changed.

Like me, Siegenfeld was surprised by the results of the Brexit referendum. But for him, it was an ocean away. It would take an event far closer to home to really get his attention. Five months after the Brexit vote, citizens of the United States went to the polls to elect their next president. Almost every pundit predicted a resounding victory for Democratic nominee Hillary Clinton. But, like the Brexit polls, they were all proven wrong. With the surprise election of Donald Trump, the world had experienced a second totally unexpected, seemingly unpredictable election outcome.

Much has been written in the years since about the social, political and economic parallels between Brexit and the 2016 presidential election. The debates around both centred on issues of immigration, sovereignty and nativism. They both represented victories for radical right-wing populist movements. They both tapped into an audience that for years had harboured resentment and anger against traditional politics and the established orthodoxy. But, perhaps most interestingly, both events saw incredibly high levels of regret.

Regret may feel like a highly subjective emotion, but it is actually something scientists have been trying to measure and quantify for hundreds of years. In the 18th century, the English statistician, philosopher and Presbyterian minister Thomas Bayes even put his name to a formula that attempted to measure regret. For any decision, he defined 'Bayesian regret' as the difference between the utility of the best possible solution and the expected utility of the actual solution. Utility is another way of describing the value or benefits of any outcome.

To give a simple example, imagine you went to a grocery store with the aim of spending $100 on nothing but healthy food. We can describe this outcome as having a utility of 100. In reality though, if you're anything like me, you may end up leaving the shop having spent only $70 on nutritious items, with the rest on KitKats. The actual utility of this solution could therefore be said to be only 70. In this very simple scenario, the Bayesian regret is defined as the difference between these two numbers, or $100 - 70 = 30$.

This is an example of an individual decision, but the same logic can be applied to the collective decisions we make as a society, even those as large as national elections and referendums. For every voter, you could, in theory, assign a personal utility for every potential candidate. In other words, if that candidate won, how much utility would they deliver to that voter's life? We might quibble over exactly how we might define or measure that utility, but we can say that for any given election, it would technically be possible to identify the optimal outcome. This would be the candidate who, when all the votes are counted, would create the highest possible total utility for society. They are the electoral equivalent of a basket full of healthy food.

In reality though, it is often the candidate with a much lower social utility who wins. The populace turn up their noses at the basket full of nutritious food and opt for the one full of processed snacks and microwaveable ready meals. This is not just because people have different ideas of what is best for them, but also because of factors like misinformation, emotional appeals and short-term thinking. Voters may choose candidates who promise immediate benefits or resonate emotionally, even if those choices do not align with their long-term best interests.

In an ideal world, the Bayesian regret after an election would be as small as possible. But the feelings of regret after the Brexit referendum were immense. While total voter turnout was relatively high at just over 72 per cent, there were large disparities between age groups. Older people, who more generally favoured leaving the EU, turned out in droves. Over 90 per cent of those over 60 voted. Younger people on the other hand, who were generally much more supportive of Remain, stayed at home in far larger numbers. Had more 18–34-year-olds in particular not abstained, the result would certainly have been different.

But this was not the only regret. In the months and years following the referendum, many of those who did actually vote Leave also felt regret. In fact, in 2022, a YouGov poll revealed that one in five of those who voted to Leave regretted their decision and wished instead that they had voted the other way. Many of these people, convinced that Remain was going to win, had put in a 'protest vote' for Leave, which they later went on to regret.

In 2023, another YouGov poll found that of the 632 electoral constituencies in the UK, only one – the seaside haven of Boston and Skegness – still thought Brexit had been the right decision. The same poll showed that 54 per cent of people felt the UK was wrong to leave the EU versus 28 per cent of people who thought it was the right decision. If Thomas Bayes had been alive in 2016, he would have measured incredibly high levels of regret. And to make matters worse for those experiencing such regret, unlike normal elections, there was no opportunity to vote again in another four or five years. The referendum was a permanent and perhaps irreversible change to how Britain worked. The British public was in the same state as the instigator of a dramatic romantic break-up who is beginning to have second thoughts about getting back together.

We saw a similar story in the United States after the presidential election. In a 2017 poll, 20 per cent of those who voted for Donald Trump said that they regretted their decision and 24 per cent said that they would likely vote for someone else in 2020.

Siegenfeld found that, regardless of whether Clinton or Trump had won, the result would have left large portions of the population feeling dissatisfied with the outcome. Some level of regret is a normal part of any democratic contest. But when feelings of regret and dissatisfaction are widespread and long-lasting, it may be time to worry.

Siegenfeld wanted to understand the fundamental nature of elections and what was causing this democratic instability. For a brief time, he even thought about quitting his PhD, but he didn't want to leave the world of physics. So with no formal background in political theory, he decided instead to embrace the subject he knew best. He teamed up with Yaneer Bar-Yam, the president of the New England Complex Systems Institute, and decided to refocus his doctoral thesis – and his career – on a new topic: the physics of democratic instability.

* * *

As two physicists, Siegenfeld and Bar-Yam's approach was guided by the scientific method. And one of their first observations was that elections are incredibly complicated. They involve the behaviour of tens of millions of people, each with their individual personalities, values, beliefs, interests and experiences. To accurately understand how each of these individuals might behave and interact with each other, let alone predict an election's overall result, would be near impossible. In chapter two we talked about how an apparently solid and uniform cricket ball is actually made up of countless subatomic particles. Likewise, an election is not an entity in itself; it is composed of the behaviour of the voters that participate in it. And even if you could fully capture this at any given moment in time, people change, and so too do their voting patterns.

But in the world of physics, it's not always necessary to know all the details of the underlying objects or mechanisms to be able to produce useful and meaningful results. Siegenfeld compares the way physicists were able to describe the behaviour of sound waves – which are essentially the aggregate motions of atoms – with great precision, long before they knew atoms even existed. 'When we apply physics to understanding the fundamental particles of our universe, we don't actually know the underlying details of the theories,' he says. 'Yet we can still make incredibly accurate predictions.'

Similarly, he says, researchers don't need to understand the motives and opinions of individual voters to be able to carry out a meaningful analysis. 'Understanding the collective behaviour of social systems can benefit from methods and concepts from physics, not because humans are similar to electrons, but because certain large-scale behaviours can be understood without an understanding of the small-scale details.'

Siegenfeld and Bar-Yam took, as a good physicist might, a first-principles approach, asking a very simple question: What are elections? What is their purpose? Writing in their 2020 paper, they hypothesised that elections are 'fundamentally, a means of aggregating many opinions into one – those of the citizens into that of the elected official'. They acknowledge that the word 'opinions' is very subjective, but for the sake of simplicity they presumed that these opinions exist on a one-dimensional continuum – from left to right, liberal to conservative, Remain to Leave. If we accept this definition, how could we then measure the success of elections?

It's reasonable to assume that, regardless of who wins, a successful democratic election is one where the final result reflects the opinions of as many people in society as possible. To test this hypothesis, Siegenfeld and Bar-Yam looked at the results of US presidential elections going all the way back to the end of the Second World War. Over a period of over 70 years, they compared the composition of the voting electorate with the election results. They wanted to understand to what extent the winning candidate represented the opinions of the population at large.

What they found was that between around 1945 and 1970, there was a strong correlation between the opinions of the electorate, as measured by extensive polling data and historical records, and the ultimate outcomes. In other words, the winning candidate – whether Roosevelt or Truman or Eisenhower or Kennedy or Nixon – was the person most people in the country at the time wanted.

What's more, despite the undoubted political, social, cultural and economic changes the country was going through during this 25-year period, the country did not see massive swings in election outcomes. Successive elections produced different governments, but they were relatively similar to each other. For example, the transition from Roosevelt to Truman to Eisenhower represented a continuity in policy focus on post-war recovery and the containment of Communism, rather than drastic ideological shifts. Elections exhibited the qualities of a stable system. In other words, they behaved a lot like the ball in the valley I described at the start of this chapter. Regardless of how political candidates were 'nudged', their natural tendency was to converge on a central position.

For anyone living in the United States in the 1970s, it might have been natural to expect this trend of democratic stability to continue forever. When making decisions in life, there's often an implicit assumption of linearity. We assume that if something has stayed the same for a long time, it will continue to do so indefinitely. And if it does change, we assume that that change will be gradual and proportional. But that's not what happened.

In the 1970s, election results in the United States started to change. The country began experiencing unexpected and dramatic election shifts. From Jimmy Carter to Ronald Reagan. From George H. W. Bush to Bill Clinton. From George W. Bush to Barack Obama. And of course from Obama to Donald Trump. Like a pendulum somehow swinging from side to side with increasing force, each of these presidents was wildly different.

You might think that the reason why the country started electing such different presidents was because the opinions of the population were also changing during this period. However, Siegenfeld and Bar-Yam argue that the magnitude of changes in public opinion alone was not sufficient to account for the dramatically different election results. Their analysis indicated that relatively small shifts in voter preferences led to disproportionately large changes in election outcomes. In other words, much like the ball perched on top of the hill, where even the gentlest nudge could send the ball careering down the slope, the system itself had become unstable.

So, the next question we must ask ourselves is, what happened in the 1970s? How did the whole system change from being stable to unstable? Physicists call changes from one state to another phase transitions. While this may sound like something out of *Star Trek*, we see phase transitions every day. When you boil a kettle to make a cup of tea, you are creating a phase transition. A 'phase' is a state of matter – in this case, solid, liquid or gas. Within any given phase, changes tend to occur smoothly and predictably. A 'transition' is what happens when there is a discontinuity and that smooth and predictable behaviour gets interrupted. It is usually brought about by a very small change in the external environment; in the case of water, it's when the temperature passes boiling point.

The water in the kettle starts out as a liquid made up of molecules of H_2O. When you switch the kettle on, the energy of the H_2O molecules slowly starts to increase. The water gets hotter. Soon it will start to bubble and become too hot to touch. But it's still a liquid. Eventually though, something changes. A tipping point is reached – a change of perhaps half a degree. At 100 degrees Celsius the water boils and becomes a gas that we call steam. Once it crosses that point, it behaves totally differently to its liquid form. The individual molecules have exactly the same chemical composition as before, but collectively they behave very differently. The point is that just a small temperature change – the change from just below 100 degrees Celsius to just above – produces this widespread change. The entire system suddenly changes from one phase to another and behaves totally differently.

We see similar phase transitions in the metals cobalt, nickel and iron. At room temperature, these metals, which are known as ferromagnets, exist in a phase of magnetism. If you heat up a ferromagnet, it will become hot to the touch and may even start to glow. But it will stay in the same phase, not losing its magnetism. That is until you heat it to one very specific temperature, which in the case of iron is 769 degrees Celsius – roughly the same temperature as a candle flame. At this temperature, the iron abruptly changes. It loses all its magnetism. You might be forgiven for thinking that the heat had somehow permanently changed the metal. But if you cool it down below that same magic temperature of 769 degrees Celsius it will, just as abruptly, become magnetic again.

It is known as the Ising effect. Like all materials, ferromagnets are made of small subatomic particles. Much like a coin can either face heads or tails up, these particles can be said to 'point' either up or down. The technical term physicists use to describe this is spin. At regular temperatures, the subatomic particles in a ferromagnet are more likely than not to spin in the same direction. In fact, it's this broad uniformity of direction that makes metal magnetic. But when you heat the metal, the spins start to change. They become disordered, pointing up or down at random. As the temperature increases, the randomness increases, but the essential phase of the metal – its magnetism – remains unchanged. That is until it reaches that magic temperature. At 769 degrees Celsius, the metal suddenly and instantly transitions from being magnetic to non-magnetic. As the metal cools down, alignment – and magnetism – will emerge almost immediately.

We can now begin to see the parallels between these ideas from physics and politics. Between the 1940s and the 1970s, US presidential elections were in a stable phase. But during this time, unseen but powerful forces were building up. Then, sometime around the 1970s, something fundamentally changed. At this point, Siegenfeld and Bar-Yam believe that United States presidential elections underwent a phase transition that made the entire system unstable.

In the case of the water or a ferromagnet, phase transitions are triggered by heat. Has someone been heating up voters? Well, in a way, yes. Siegenfeld and Bar-Yam argue that the increasing polarisation of the candidates created a vicious cycle of increasingly 'heated' elections. One way to measure this metaphorical heating is to look at the actual language used during campaigns. They measured the fraction of polarising words used during each election. Between 1940 and 1970, the level of polarisation was relatively flat and stable. But from the 1970s we start to see a massive and increasing divergence.

One of the causes of this increasing polarisation was actually the way in which candidates for presidential elections were chosen. For much of the 20th century, while a handful of states held non-binding primary elections to gauge popular opinion and guide the selection of candidates before a national election, the Democratic and Republican parties formally chose their candidates in private meetings behind closed doors in smoke-filled rooms. Without facing a popular vote, this process was understandably considered too opaque and undemocratic. The weakness in the system was famously highlighted in the 1968 presidential election when the then vice president, Hubert Humphrey, was nominated as the Democrat candidate despite not having won a single election. He lost, badly, to the Republican Nixon. Since the 1972 election – Nixon's second – candidates are now more likely to be chosen based on their performance in primaries. It is no coincidence that the era of increased polarisation begins with Nixon.

The effect of shifting more to primaries, you would think, would be to produce more candidates who were truly in line with the preferences of the electorate. On the face of it, this does seem the more democratic option. However, what actually happened is that the candidates chosen were the ones good at winning primary elections, not necessarily those best suited for a national election. They were more likely to have been selected based on their response to partisan issues, without reference to the broader view that is required of a national leader. The meetings behind closed doors might have felt like a relic from a less democratic age, but they arguably allowed more strategic selection of the candidate best suited for the national vote.

This polarisation has a dramatic effect on the electorate. A simple way to visualise this is to imagine a simple graph with political preferences represented from left to right and the number of people who held the political view associated with that position from bottom to top. From the 1940s to the 1970s, the shape of this graph resembled a bell curve. There is a peak in the middle, gradually tailing off as you get further towards the edges on either side. As we learned in chapter one, bell curves are common in physics.

For elections, what this bell curve means is that most people's opinions lay somewhere in the middle of the political spectrum. It's natural therefore to see why any potential candidate, to maximise their chances of winning, would aim for the centre, because that's where they stand the best chance of reaching the most voters. If voter preferences changed, they would move to the left or to the right of centre but – because of the high concentration of voters in the middle – the changes would not be dramatic. This created an inherent stability in the system, like the balls in the valley.

But when the electorate becomes more polarised, you no longer see a smooth bell curve. As more and more people start taking more and more extreme views, you start to see a very different shape. Fewer people hold views towards the middle of the political spectrum and so this part of the graph dips down. And in its place, two peaks emerge towards the left and the right. In physics this is called an M-shaped graph.

If you were a politician trying to win an election faced with an M-shaped electorate, what strategy would you adopt? It would no longer make sense to go for voters in the middle. You would be far more likely to take a position that appeals to voters in one of the two peaks, ignoring the views of the voters on the other side. There is no force driving all parties towards the centre.

What this means is that when there are small changes in the electorate's opinions, you no longer get a proportionately small change in outcome, but instead a potentially massive swing. And with each successive election, political parties start to nominate candidates further from the centre. Thus there is a vicious circle of polarised voters leading to polarised candidates leading to more polarised voters.

And this leads us to perhaps one of our most surprising findings. Imagine again our ball perched on top of a hill. We know now that this is an unstable system. If we were to give the ball a little nudge to the left, you would right-fully expect it to roll all the way down the left side of the hill. But imagine if it actually rolled all the way down the right side of the hill. That would be totally count-er-intuitive, yet that is exactly one of the implications of Siegenfeld and Bar-Yam's work. It's called negative representation.[1]

They found that whenever you have an unstable elec-tion or make any collective decision in the context of an unstable environment, there will always be some opin-ions that are negatively represented. What that means is that if the political opinions of the citizens move in one direction, it actually perversely causes the outcome of the election to move in the opposite direction. And this is exactly what happened in 2016 both in the US and the UK.

In the case of the presidential election, voters were faced with a choice between a centre-left candidate, Hillary Clinton, and a much further-right candidate, Donald Trump. In a stable system, the more left-leaning voters there are, the more likely it is for the centre-left candidate

---

[1] 'Negative representation and instability in democratic elec-tions', Alexander F. Siegenfeld and Yaneer Bar-Yam, *Nature Physics*, 16, February 2020. https://www.nature.com/articles/s41567-019-0739-6.epdf.

to win. But if the overall sentiments of the electorate move too far to the left, those voters might regard both Clinton and Trump as equally unpalatable and so they decide not to vote at all, or vote for a third-party candidate such as Jill Stein. They stay at home on election day and therefore they reduce the left-wing vote so much that there is no balancing force towards the centre. The reduction of the far-left vote has led in fact to a stronger vote for the candidate they dislike the most. The far-right candidate ends up winning, even though that's not what most people in the country want.

Similarly with Brexit, voters faced a choice between the highly emotive right-wing rhetoric of the Leave campaign and the moderate, less charged arguments of Remain. Perhaps thinking that no one could possibly swallow the claims of the Leave campaign, many Remain-leaning voters, especially younger ones, unexpectedly stayed at home, helping tip the result. And of course this is not just about Donald Trump or the campaign to leave the EU. The same principle would apply if right-wing voters decided to eschew both a centrist candidate and a far-left candidate. With negative representation, a shift in the electorate's opinions one way actually shifts the outcome of the election in the other direction.

* * *

In short, instability leads us to make bad collective decisions as a society, which is in turn bad for democracy. Ideas from physics can, however, offer a potential path to greater stability.

One of the most immediate ways to increase stability is by increasing the number of people taking part. As we saw after the Brexit referendum, not everyone who is eligible to vote, votes. This may be because they are too busy or distracted. Or it could be because of apathy, or complacency, or simple disillusionment with the whole process. There may be many convinced their vote won't make any difference, so why bother?

One might think that lower turnout would simply reduce the outcome of the vote proportionately. But low turnout exacerbates the M-shape voter demographic we saw above. It forces candidates to spend less time focused on swing voters in the middle, and even more time drumming up support from their core base.

For some people, it is also tempting to think that it doesn't matter if many citizens don't vote because most people are very uninformed and so it is best to keep their ignorance out of the public field and leave the voting to the grown-ups. In fact, research shows that is wrong. Regardless of how informed or uninformed voters are, high voter turnout will contribute to stability, while low voter turnout will result in instability and negative representation.

So the lesson is: Even if you dislike all the options available, swallow your pride – possibly even hold your nose – and vote. The more people who participate in a decision, the more stable the system and the more likely the outcome is to reflect public opinion. Widening the voter base reduces polarisation. If you imagine our balls in the valley, you can even visualise this: The larger the base of the valley, the deeper and more bowl-like it becomes, and the more the balls are pushed together. The smaller the base, the easier it is for them to fall off the edge. If you vote, then not only are you making your own voice heard, but you're also doing your bit to contribute to the long-term stability of our democracy.

Beyond suggesting more people should vote, physics offers a deeper lesson. Current political institutions are built for a classical, Newtonian world. They are not equipped to deal with a world that is, as we learned from Sarkissian earlier in this chapter, increasingly quantum. Take politicians, for example. They tend to represent a binary view of the world: yes or no, black or white. After all, they are either in office or they are not, and a vote not for them is against them. However, real politics tends to be more quantum, requiring us all to embrace multiple and even contradictory points of view at once.

Consider a voter who has always regarded themselves as generally centrist, and in principle supports higher taxes to pay for public services. However, they also run a struggling business that is slowly sinking beneath its tax burden. At the election, they must choose between two parties: a right-wing party that favours low taxes, or a left-wing party that wants to raise them. Current voting models force this voter to make a binary choice, even though ideally they would like the new government to give each idea its due. The voter wants to know that both of those pressures are being weighed equally and that a decision will be made with the greatest social utility, rather than due to rigid ideology. But there's no room for this nuanced 'quantum' thinking.

To return to the polarising subject of Brexit, while there were voters who either uncritically supported or entirely loathed the EU, most voters were prepared to accept that it had both strengths and weaknesses. However, the binary 'classical physics' nature of the poll was unforgiving of such ambiguity. Voters had to choose one or the other.

One take-home lesson here is that referendums are a very bad way of making national decisions, especially when the threshold is set at 50 per cent – 51 per cent of voters get exactly what they want, 49 per cent get nothing. I will always regret the result of the Brexit referendum. To me, it was an unnecessary push that finally broke up a volatile relationship, leaving both sides with resentment, hurt and a lingering feeling that they were the injured party. Perhaps each side took the other for granted while the relationship lasted, and let tensions grow with no effort to resolve them. Even the most stable relationship is a dynamic thing, in need of constant care and attention. Brexit showed up so many flaws on both sides that there will be no quick kiss and make-up. If we do rekindle our relationship, it will be a long way into the future and take a lot of work by both parties. But, like many disappointed lovers, we can learn the lessons from this time around to make what happens in the future truly lasting.

As for me, after my history of painful break-ups, I have thankfully been able to find love and happiness in a stable relationship. At the time of writing this, I have been married for 13 years. While I certainly can't claim to be the best husband or to have a perfect relationship, ideas from physics have proved useful. In physics, certain critical points, like a specific temperature, mark the thresholds where substances undergo significant changes. And just as substances undergo such phase transitions – from solid to liquid to gas – so too can relationships. Over the course of our relationship, my wife and I have experienced many such transitions – from dating to living together, to getting married, to moving home, to becoming parents, to losing loved ones. Passing through each of these thresholds in our relationship has made our bond more resilient. Stable does not mean unchanging; rather, it means successfully navigating these transitions, surviving and emerging stronger. Change is inevitable. But the way we handle these transitions is what fosters true stability.

Chapter 6

THE PHYSICS OF NOT FITTING IN

If you were visiting my home and I offered you tea, what would you expect? A cup of boiling water infused with the flavour of dried leaves (perhaps accompanied by a small plate of chocolate biscuits)? Or a sit-down evening meal? Alternatively, if we met in the street and I invited you for dinner, would you turn up around midday or in the early evening? If you live in the British Isles and tea, for you, is the hot drink, or dinner is an evening meal, then – like me – you probably come from the south. If you identify more with the alternative definitions then – like my wife – you're probably a northerner.

I was born and bred in London (latitude 51.5° N). For my higher education I ventured as far north as Oxford (51.7° N), and after graduation snapped straight back to London. Despite that, I met, fell in love with and married a woman from York (over two degrees further north). She and the rest of the family share accents with the likes of actor Sean Bean and Spice Girl Melanie B. I have a vivid memory of my father-in-law, in his speech at our wedding, drawing attention to this fact but assuring me I would eventually fit in . . . It would just take about 50 years, a common joke among those from Yorkshire towards newcomers and incomers.

Of course, to belong somewhere, you actually have to be there. Our first married home was in London, but the transition began when, not long after we were married and before we started a family, my wife and mother-in-law convinced me to move up to York. The 200-plus-mile commute to work in London every week took some getting used to, but being near family and having glorious scenery on our doorstep have made the move more than worth it.

We recently marked our tenth anniversary of living up north, but I am still getting used to the different ways of communicating. My native northern children speak in a proper Yorkshire accent and tease my pronunciation. I pronounce the green stuff that grows on our lawn 'gr-arse', or 'grāss' according to the phonetic alphabet. They say it as if the 'a' is barely there: 'grăss'. But at least it is still the same word for the same thing. Then there are words like 'tea', common to both north and south but having subtly different meanings. And then there are the words that are unique to the area. In the south, kegs are what you store beer in; in Yorkshire, they're another word for trousers.

Combine a regional accent, local usages and unique words together, and you have a dialect. It is estimated that over 7,000 different languages are spoken around the world, and there are countless more dialects and variants within these. There are 30 variations of English within England alone, and I travel through many of them as I take the train up and down the country between York and London every week. Each of these accents and dialects is a variant in its own right with a unique culture and heritage.

*　　*　　*

Accents and dialects were once solely the domain of linguists, but now British physicist James Burridge, of the University of Portsmouth's School of Mathematics and Physics, has become interested in their evolution, variation and flux.[1] He is a physicist who has spent most of his life applying physics to things seemingly unrelated to physics. His interests include naturally occurring rock formations

---

[1] 'Spatial Evolution of Human Dialects', James Burridge, *Physical Review X*, 7, 17 July 2017. https://journals.aps.org/prx/abstract/ 10.1103/PhysRevX.7.031008.

and the dynamics of how games play out – he is interested in looking for ways to make predictions about things that are seemingly unpredictable. But he has a particular interest in language and dialects, and the use of physics to model them. His inspiration: bubble baths (pronounced 'bāths' or 'băths' – your choice).

Bubbles are governed by the laws of fluid dynamics, the branch of physics related to the behaviour of liquids and gases. In the 18th century, the Swiss mathematician and physicist Leonhard Euler believed that a set of laws could be developed to explain the behaviour of fluid based on Isaac Newton's deterministic laws of motion. In other words, if we know the initial state of a fluid system and the laws that govern its behaviour, we should be able to predict its future entirely. Euler's equations, however, were not able to take into account the fact that fluids in the real world have viscosity and exhibit friction.

Two individuals independently took up this challenge. In 1822, Claude-Louis Navier, a French engineer and physicist, made the first attempt. He observed how fluids flow through pipes and developed a set of equations that represented viscous stress. A few years later, George Gabriel Stokes, an Irish physicist and mathematician, published similar equations. Unlike Navier, he derived his equations from a microscopic perspective, considering the interactions between individual fluid particles. Navier's and

Stokes's work did not gain prominence for many years. But later in the 19th century, experimental investigations into fluid mechanics confirmed the accuracy of their work. The Navier–Stokes equations now represent a set of complex rules that encapsulate the dynamic behaviour of fluids.

Like the laws of motion and thermodynamics that we've already looked at, they are predicated on the conservation of three things: mass, momentum and energy. Conservation of mass means that the mass of a fluid system remains constant over time. Conservation of momentum, which is defined as mass multiplied by velocity, means that the total momentum of a fluid system must remain constant over time, unless acted upon by an external force. And conservation of energy states that the total energy of a fluid system remains constant over time, unless acted upon by an external force.

The laws of fluid dynamics can be used to describe a wide variety of phenomena, including the flow of water in rivers and pipes, the flight of aeroplanes as air passes over the wings, the movement of air masses in the atmosphere . . . and the behaviour of soapy bubbles in a bubble bath.

All bubbles are formed by surface tension. One implication of the conservation laws described above is that they try to make their surface area as small as possible. The 3D shape with the lowest surface area is a sphere, so that's why a solitary bubble on its own is spherical.

Bubble baths and dialects might sound like they have absolutely nothing in common, but Burridge found a link. He argues that the same physics that describes bubbles can offer new insights into the formation of language patterns. The boundaries of a bubble, Burridge believes, are analogous to isoglosses – lines drawn on a map to show the boundaries of the regions where different language components change. Just as bubbles seek to keep their surface area as small as possible, the shape and size of these linguistic boundaries adjust to the dynamics of social interactions, population movements and geographical features.

You can picture this by imagining a map of a country, but instead of cities and rivers, you see a cluster of bubbles. Each bubble represents a different dialect region. These bubbles aren't perfectly round; they're squashed and squeezed by their neighbours, forming irregular shapes and intricate patterns. The country is a container and these dialect bubbles are jostling for space, constantly pushing and reshaping their boundaries. Some are big and dominant while others are small and squeezed. The borders of the country act like the sides of a jar (albeit a jar with a very irregular shape), containing the bubbles and forcing them into the most efficient arrangement.

Burridge's insight was to identify how surface tension, combined with other factors like variations in population density and the shape of the country, produces the patterns we see. He found that the same physics that describes bubbles can predict what dialects we hear, and where and when.

It can also predict how language will change. Factors like population movement and migration can cause small linguistic bubbles to merge into bigger ones, and Burridge believes that this model may allow us to predict the dialects of the future. Or, in many cases, the end of these dialects. The implications of Burridge's work paint a stark picture. Over the next 50 years, it is predicted that many regional dialects will become increasingly alike to their neighbours, if not entirely obsolete.

In some ways, this is unsurprising and not especially alarming. We all know that language changes. The English of this book is very different to the English of Geoffrey Chaucer, 800 years ago – or even of Shakespeare, who wrote in the middle of that time gap. Loss of languages is not like the existential threat posed by risks like climate change or economic inequality. It was not one of the topics discussed at Davos. But it can represent the loss of something important.

There are many reasons relating to culture and identity why people may want to preserve their dialects. Science shows that, fascinatingly, there is also good reason to believe that diversity of dialects is closely related to biodiversity – the numbers and variety of animals, plants, insects, fungi and other life forms coexisting in one area. Areas with more diversity in dialects are also more biodiverse, and where the numbers of dialects are dwindling, so are the numbers of species.

Biologists currently estimate the annual loss of species to be at least 1,000 times greater than the historical rates observed over the Earth's history. At the same time, linguists expect that between 50 and 90 per cent of the world's languages will have disappeared by the end of this century. In other words, languages and species are dying together. Most tellingly, these losses are happening in the same areas.

The world has 35 designated biodiversity hot spots – regions with exceptionally high occurrences of endemic species, i.e. species that live in a limited area, like a mountain range or a particular lake. Those 35 hot spots contain 3,202 of the more than 6,900 spoken languages around the globe – nearly half. There are also five designated biodiversity wilderness areas – regions of at least 10,000 km^2 with relatively little human impact that have lost only 30 per cent or less of their natural habitat – and they contain 1,622 different languages.

An endangered language can be classified as one with 10,000 or fewer speakers, which describes 1,553 of the languages in biodiversity hot spots. Of those, 544 have fewer than 1,000 speakers. It is also these hot spots and wilderness areas that are seeing the greatest loss of biodiversity.

Perhaps this is inevitable. As a rule, the greater the population in a region, the smaller the biodiversity. Compare the wilderness that was the east coast of North America in the 16th century, before the influx of European settlers, to what it is now. Larger populations tend to homogenise their environment in order to make living there easier. That means eliminating – deliberately or otherwise – species or plants that are inessential for or hostile towards human survival. So, high biodiversity regions will almost by definition have smaller populations, and with fewer people around to begin with, it follows that any language that is spoken there will not have many speakers.

Then again, there is no obvious reason why a region with a small population should host a large number of languages in the first place, though the fact is that many do. What generates high linguistic diversity in Burridge's model is a combination of people staying local, meaning they interact more strongly with those nearby, and a highly heterogeneous population distribution, with each community being quite distinct from the others. Having more plentiful and diverse resources reduces the need for local populations to communicate and cooperate with each other; hence, there is no need to regularise their languages.

Professor Larry J. Gorenflo of Penn State University argues that local languages are strongly linked to local culture, and that culture enables the preservation of biodiversity. People living in biodiversity hot spots and wilderness areas tend to be knowledgeable about the plant and animal life around them. They tend to have traditional practices that include, either deliberately or as a by-product, conservation of the local environment. Conservation in these areas often benefits from the traditional practices of the human inhabitants, including sustainable agriculture, controlled burns to manage forest health, and sacred groves where wildlife is protected, which can improve biodiversity beyond what it might be if the areas were uninhabited. The survival of the inhabitants is intricately linked to the preservation of their habitats. The loss of linguistic diversity occurs when small groups who have traditional information about rare and fragile environments are lost. These tend to be replaced by modern industrial economies that require a more simplified environment. So, preserving the people preserves the habitats; preserving the habitats preserves the people.

In short, the drivers that link linguistic diversity and biodiversity in the first place seem to be complex – but the fact that there is a link is clear. To understand how ideas from physics can help, let's take a closer look at how human language actually works.

* * *

Language consists of a lot of different components. There are not only the words themselves (the lexicon), but the single units of sounds (phones) that combine to make them; the rules for combining those sounds (phonology); the rules for how words are then constructed (morphology); and the rules for how those words are made into sentences (syntax). Languages change as their components change. Either there is a slow mutation and evolution, with words changing their meaning slowly over time, or new words come right in and replace the old more abruptly.

At any moment there might be variants that are still recognisable as the same language, and this brings us into the realm of dialects. Sometimes it's just pronunciation, like the aforesaid 'grāss' and 'grăss'. Sometimes different words are used for the same thing (the female possessive pronoun is mostly 'her' throughout the UK, but a distinct variant, 'hern', can still be heard in Yorkshire), or the same word is used for different things (the tea/dinner divide, for example). Sometimes sentences are structured differently: In different places around the UK, the time that is 8:45 in the morning might be referred to as a quarter to nine, or a quarter before nine, or a quarter until nine. And sometimes there are distinct quirks, like the Irish habit of adding 'so it is' to the end of a sentence to emphasise its truth, or the Midlands habit of addressing anyone, regardless of gender (or indeed species), as 'duck'.

Languages also require speakers. It is the speakers themselves who define how language changes. It is very hard to follow or predict the actions of speakers as individuals. But put them together, and researchers find themselves able to track long-term, macro-scale patterns of use, and observe universal laws in how languages act.

As James Burridge puts it, language is a good place to apply physics because it satisfies some of the conditions that make statistical physics work. For a start, it is created by lots of interacting units – i.e. people. However, lots of other things are also created by people: political systems and all sorts of messy bits of history. For physics to work, you also need to be able to generate macroscopic equations based on microscopic assumptions about individual behaviour. The macroscopic behaviour needs to be a result only of those small-scale interactions, and not the one-off action of some individual or some massive external force, like technological change or a devastating plague. Language is a good candidate for a physics-type model because it fits that description nicely, being generated by lots of microscopic interactions, with only a few more macroscopic effects, such as American films exporting US terminology around the globe, the influence of colonialism on local languages, or the establishment of received pronunciation as a standard in media and education. Each of these has had a profound impact on how English is spoken and understood worldwide.

In this way, physicists have been able to create a new interdisciplinary field: language dynamics.

One of the most interesting studies demonstrating the physics properties of language was carried out in Japan. The 2011 study looked at how 21 swear words spread across Japan. Swear words are a good candidate for study, because while they follow the same rules as the rest of language, they also – especially in a culture that prizes politeness, like Japan – tend to stand out.

Japan's main islands give the country a long, thin shape, running south-west to north-east, so from a physicist's point of view it is almost one-dimensional. Picture rainwater trickling over the surface of a road. When it comes to flat, open areas (two-dimensional planes) it tends to pool and swirl around, and slow down. When it comes to a narrow, one-dimensional channel in the tarmac, it surges ahead. Travel in one dimension forces faster dispersal, giving less time for a gradual mutation. The result is that the long, thin shape of Japan means that new words move fast.

The study looked at linguistic maps that showed the use of swear words. Most began in Kyoto, a former capital of Japan that is often regarded as the country's cultural centre. Kyoto is almost central to Japan's geography – it sits about halfway between the north and south coasts, at almost the narrowest point of Japan. It is also situated a little under half the distance from the south-western to the north-eastern tips of the island.

To picture how swear words spread up and down the country, picture concentric circles emanating out of Kyoto, forming wavefronts that move over land, heading north-east and south-west. Most of the words were found spread out along wavefronts at similar distances in either direction from Kyoto. So, if a word had spread 100 miles south-west, it had also spread 100 miles north-east. The team worked on the assumption that there is a regular stream of linguistic innovations, originating in Kyoto. These spread out across the country, meeting older terms and possibly replacing them. For example, a foolish person would probably be called 'aho' ('dumb') in Kyoto but 'baka' ('stupid person') in Tokyo. 'Baka' also used to be predominant in Kyoto, but at some point it was overrun by 'aho'. But at the time of the study, 'aho' had not yet reached Tokyo. The newer the word, the less far it had travelled from Kyoto.

Starting from just a few examples and working on these assumptions, the researchers were able to accurately model the actual distributions of all 21 words – and predict the distances between the wavefronts. They showed that language was following the rules of physics.

James Burridge began his work, not with swear words or even humans, but with birds, specifically Puget Sound white-crowned sparrows. Many animals and birds have

the equivalent of dialects, in the forms of variants of a particular sound or squawk or song that you find in different
areas. Birdsong has the advantage of bringing the same
principles as human speech to the table, but in a far simpler
form. Human speech has plenty of variable elements –
pronunciation, phonology, syntax – and all of them can
have variants. While it's possible that birdsong contains
similarly distinct components that we do not yet understand, it has one dominant element that we perceive as the
song itself. The relative simplicity of birdsong compared
to human language makes it an ideal subject for studying
the basic principles of dialect formation and evolution. For
example, variations can be found in the trills of male sparrows over different areas, shifting and changing as different
populations merge with migration.

The Puget Sound white-crowned sparrows also have
the advantage of breeding grounds that are quasi-one-
dimensional, like the islands of Japan – long and thin,
without the distraction of too much sideways movement.
The general area of the breeding ground of the sparrows is
the Pacific Northwest coast of the US, but it is divided into
distinct regions. All the males sing roughly the same song in
each region, but when they cross the regional boundaries,
the dominant song of the males completely changes.

Drawing on spatial maps of birdsong created by ecologists, Burridge was able to see how the patterns changed

over the life cycle of birds. From the moment they hatch, young male birds enthusiastically fly around different areas, learning a repertoire of many different songs. Older birds on the other hand generally travel less far, with the result that they tend to sing the one song that is dominant in their region.

Things get interesting when winter comes. As you would expect, all the males fly away to overwinter somewhere warmer. Some survive the winter, and some die. The older males then return home to their area before the younger ones. They settle in and their song once more becomes dominant. Eventually, the younger males return, and they must fit into the vacancies created by the deaths of older birds. Even though their songs are more varied than those of their elders, to show reproductive desirability to the females, they have to learn to copy the song of the older birds. It is therefore in their interests to copy their neighbours. They quietly forget the varied playlist of their youth and learn to sing just one song. That is the song they retain in later years. They then become older birds themselves. A new generation of younger birds is born, and the cycle repeats.

Replace the words 'song' with 'dialect', and you begin to see Burridge's theory of language. Speakers mostly interact with people living in their local area, and hence their speech patterns conform to the majority speech patterns in that area. 'Local area' is a relative term, but typically it is between a few miles and a few tens of miles wide, and centred on the speaker's home. We collect memories of the speech patterns around us and they then affect our own speech behaviour. However, memory decays, so it is always the most recent memories that dominate our behaviour. The result is that where a speaker finds themselves in a minority, they will adjust their speech pattern – consciously or unconsciously – to match the current majority.

This makes intuitive sense; it is far more likely that a solitary newcomer to an area will adjust to speaking more like the established majority, rather than vice versa. If you are simply passing through somewhere, not lingering for any length of time, then you will probably not have an effect on other people's speech and they will probably not affect yours. But if you stay in that area long enough to acquire neighbours, then you will begin to conform to the patterns of speech all around you.

However, Burridge's research with the birds went beyond a simple statement of principle about how dialects emerge and disappear. He began to see the physics. By studying maps of the different songs, he found that dialects disappeared if the death rate of the birds went above a certain level. If the death rate fell and the competition survived, the dialects began appearing again.

Burridge realised that these transitions mirrored another physics phenomenon. In the previous chapter, we learned how in ferromagnetic materials, each atom acts like a miniature magnet that wants to align itself with its neighbours, pointing either up or down. When all the atoms are aligned, the material has a phase transition and becomes magnetic – and it happens very suddenly. One moment there is no magnetism, the next moment there is.

With ferromagnetic materials, the external factor is temperature. The warmer the material, the more unaligned its atoms will be. Cool the material down and it abruptly becomes magnetic; warm it up again and, just as abruptly, the magnetism disappears. In the case of the sparrows that Burridge studied, the death of a male bird is analogous to the temperature of a ferromagnet. There comes a point when a region of birds has a phase transition: Suddenly all are singing the same song.

With those rules in place, Burridge began to see how physics could be applied to our understanding of language.

If you travelled around England, you might find local references to gorse, furze, whim or broom. Four different words but they are in fact all describing the same thing – a prickly shrub that likes to grow on dry, sandy soils. Imagine that the use of these four different words was marked on a map with arrows. Anyone saying 'gorse' is marked with an up arrow, anyone saying 'furze' is marked with a down arrow, 'whim' is left and 'broom' is right. What pattern would you expect? Instead of a jumble of arrows all pointing in different directions, the map will show four regions where all the arrows generally point in the same direction. Like particles in a magnet, each speaker has aligned along with their neighbour. A speaker from one region who moved into another would eventually find themselves using the same term. They would realign.

And this brings us back to bubbles. If we were to draw a line around an area where arrows are all aligning in the same direction, you get an isogloss. These isoglosses are analogous to bubbles in a container. Dialects are contained within their own bubbles. Perhaps, Burridge asked, isoglosses don't just look like bubbles, but even act like them.

Picture a mass of bubbles in a bath. Because of surface tension, over time, bubbles will either shrink to nothing or merge with others to form bigger bubbles. If you leave the bath for long enough, eventually only a few bubbles will remain, and their common boundaries will no longer be curved, but straight. The bubbles settle down in equilibrium. The forces on either side of the barrier, from within and without, are balanced.

In physics, this is called coarsening. Surface tension leads to the gradual disappearance of fine distinctions and the pattern becomes coarser and coarser. The boundaries of the shapes try to minimise their own length by straightening out when they are pressed against another region, curving when they are not.

Isoglosses also take on this appearance. Generally speaking, we find large areas where one linguistic variant is in common use. Between the larger areas are comparatively narrow transition regions, where another variant may be used more or less equally. Beyond that transition zone, we move into another larger area where this time the second variant is dominant. Whenever you approach an isogloss, you find people start to use a different variant for the same thing – perhaps in pronunciation, or choice of words. Children growing up on an isogloss are exposed to language on both sides and so are equally likely to use either term, meaning or pronunciation, as there is no particular pointer to what the 'right' choice is.

Indeed, it is often children who cause the isogloss to shift. Dialectologists who collected survey data at the start of the 20th century found almost the whole population pronounced the word 'thawing' as almost two distinct syllables, 'thaw-wing'. (We will see where this data came from later.) A variant pronunciation – 'thawring' – crept in for the simple reason that it uses less energy to say out loud and so was more readily adopted by children learning to speak. Over a period of about 25 years, this new variant became the norm . . . mostly. While it was absorbed else-where, it hit urban centres like Leeds and Manchester, and

bounced off. Those cities were sufficiently densely populated that the new pronunciation was repelled. The variant
got stuck around their edges, creating a new isogloss.

One way that isoglosses are different to bubbles, however,
is how they expand under their own power. Bubbles in a
bubble bath will expand when their mutual boundaries give
way under surface tension and the contents merge, so that
a mass of small bubbles becomes a smaller number of larger
bubbles. A bubble will expand in this way until it reaches
an obstacle that blocks it. This might be another bubble
coming the other way, or the edges of the container holding
it. Thereafter the boundary progresses at right angles to
that obstacle. The boundary settles at an equilibrium point
perpendicular to that obstacle.

So, bath bubbles will expand, but only when surrounded
by other bubbles. Left to its own devices, a bubble on its
own will only want to contract.

On that basis, an isogloss on its own should also contract,
moving in towards the centre of population. Yet isoglosses
actively want to expand, even if they are entirely on their
own. This is for the simple reason that an isogloss bubble
consists of people, and people have their own agency.
They want to move about. They grow up and leave home.
They move to new areas and they meet more people. We
are back in the realm of the microscopic interactions that
Burridge mentioned earlier.

Variety in a language starts to creep in as speakers meet speakers from other domains, but this only happens in sufficient quantity to have an effect out on the edges. The more concentrated a population, the more likely it will speak the same way. Where there is an isogloss with a denser population on one side than the other (call them town and country, for convenience), you will tend to have more conversations with town people than country people. So, you will tend to use the linguistic variants found in the town. The more people who do this, the more the isogloss is pushed outwards, away from the centre of population into the rural areas.

The result is that cities tend to lie at the centres of isogloss bubbles and create their own dialects. Think of cockney or the more recent MLE (Multicultural London English), a recognisable dialect that has emerged since the late 20th century as young working-class people from different ethnic and cultural backgrounds, not just in London but also in other cities, rub shoulders together at home and at work.

One by one, these expanding isogloss bubbles will bump into each other and merge. The bubbles grow larger, the remaining isoglosses less curved. Eventually a single strong isogloss will emerge between two more or less equally dominant dialects. Neither side will overwhelm the other. The overall tendency is for very many variations of language to settle down into just a few.

For the same reason that isoglosses expand away from population centres, they are also influenced by geographical features such as indented coastlines, bays or river mouths. Where there is an indent, the isogloss seems to move towards it. This is for the very good reason that there are people on the land side of the isogloss, but no one on the water side. Even a small population is denser than a population of zero. Once they meet a natural obstacle, then all other things being equal, isoglosses will tend to settle down at right angles to the apex of that natural obstacle. Again, they are showing the same behaviour as bubbles that reach an obstacle.

To go back to the Japanese example, remember that the waves of linguistic change expanded from Kyoto in all directions. In all but a couple of directions – north-east or south-west – they very quickly met the coast. The only wavefront that counts is the one expanding over land. Draw it on a map and, again, you have almost a straight line moving north-east and south-west, at right angles to the geographical barrier of the sea. The same effect is found in the animal kingdom. The yellow-naped Amazon parrot lives in Costa Rica, part of the narrow isthmus joining North and South America. The parrot has two distinct dialects, measured in terms of how it constructs its song, and the boundary between these runs perpendicular between the north-east Caribbean coast and the south-west Pacific coast.

In mainland Great Britain, the main isogloss between northern and southern English dialects is a straight line between two large indentations in the coast: the Severn estuary in the west and the Wash in the east. Over the centuries, many isoglosses expanded and merged into larger bubbles across the country. The isoglosses that reached the apexes of those indentations settled down at right angles, moving away from them. This eventually averaged out into two large bubbles, one in the north and one in the south, pressing together in a straight line between the two indents.

Another effect of this is that countries that are long and thin tend to follow the pattern that linguists call the Rhenish fan. This is where isoglosses fan out from a population centre, and it was first noticed in the cities along the river Rhine in Germany. Germany has one main isogloss running across the country, again showing the rule that isoglosses tend to be perpendicular to natural obstacles, in this case the Rhine itself. Low German is spoken in the north and High German in the south. However, as it approaches the western frontier of Germany, this one main isogloss fans out into at least eight different lines. Each one of these comes out of one of the Rhine's cities.

A similar effect has been found by researchers who look at the dispersal and evolution of the Numic languages spoken by Native Americans in the western US. A single proto-Numic language fanned out into multiple varieties as the speakers spread out inland. In 2006, researchers from the Universidade Federal de Pernambuco in Brazil were able to reproduce this pattern using computer simulations. Language divisions in Belgium and Catalonia – each multilingual territories with distinct clusters of monolingual populations – show the same results.

*   *   *

Burridge's theories and the assumptions they were based on have consistently proved to be correct. With them, he has been able to build a physics model that uses equations for how linguistic variables are spread over an area, mapping not just where they are now, but where they have been, and where they will end up. Take any country, describe its basic shape, factor in the varying areas of population density, and the natural indentations and obstacles around its border, and you should be able to see how the dialects within it develop.

As Burridge's model is based on the assumption of people adopting their speech patterns to those around them, it has to account for human behaviour in general. People move around in their home locations. A job might take them further afield, but then they return home; marriage might also take them further afield and there they stay. He also looked at statistics showing population distributions and migration patterns from the 1900s onwards. This kind of population data was already available from other researchers. To it, Burridge added his own assumptions of how people acquire language.

A model needs data to test itself against. Burridge used data provided by two key language surveys. The first, a Survey of English Dialects (SED), was carried out in the 1950s. The second, carried out via the English Dialects App (EDA) in the early 2010s, surveyed respondents through a smartphone app. Similar questions were asked, such as: How do you pronounce the word 'last'? ('lāst' or 'lăst'); If it belongs to a woman it's . . . ('hern'/ 'hers'); An animal that carries its house on its back is a . . . ('dod-man', 'hodmedod', 'hoddy-dod', 'hoddy-doddy', 'snail'). The results not only produced a series of isoglosses but, being conducted over 50 years apart, also let isoglosses be tracked over time.

Rolling the model forward, Burridge compared it to present-day data from dialect maps and found that it matched. His model had predicted how the English language evolved across the 20th century. Based on this, he could roll it further forward and predict how language will continue to change.

In the same way that magnetism comes to dominate a piece of metal because all its atoms align, or the bubbles in a bubble bath merge together, Burridge says that simple physics dictates that the number of dialects in a country will fall over time. The UK is not exempt. Isoglosses will fade and the bubbles will join up. With current levels of both domestic migration and emigration, once a variant is used by the majority of the population, unless there is some strong geographical feature or local cultural conformity to stop it, it will expand. Because the UK is densely populated in the south, says Burridge, it is the southern variants that will tend to expand via this mechanism. Words like 'strut' will no longer rhyme with 'foot' in the north of England. The long, somewhat piratical 'arr' in 'farm' will disappear from the south-west. Nor is it just pronunciations that will disappear. Whole words will go the same way. The possessive pronoun 'hern', mentioned earlier, was already used only by a distinct minority in the SED. The 'hern' isogloss continues to shrink and, despite rearguard defensive action in Yorkshire, it will probably disappear within 100 years.

Not every variant is set to disappear, however; according to Burridge's models, my kids will be glad to know that there will still be no regular pronunciation of 'grass' throughout the UK.

The phenomenon is not just confined to the UK, any more than physics only applies to one country. Burridge's model applies anywhere there are dialects marked out by isoglosses, and that applies to large chunks of the Earth's surface. Languages are standardising in every direction.

This does not mean that there will one day be a universal language with no dialects at all – and nor would Burridge want there to be. On the one hand, he says, a degree of conformity in behaviours and convention is good for a society. It leads to cohesion and stability. But it is also possible for stability and diversity to coexist. There are things we can do to help us feel closer to one another and cooperate more, and speaking the same language as our neighbours is just one of them. 'Speaking like other people, and knowing that you're from somewhere by the way you talk, is good. It makes places feel like places. It's the same with jobs. People want to work for an organisation where they are not just a number but have also got an identity. So, yes, language is an important part of our identities. And I think that actually having distinct identities is an important aspect of being human.'

✳ ✳ ✳

So, how does this help us with the problem that we started with: the link between declining linguistic diversity and diversity of species? We know that loss of linguistic diversity tends to indicate loss of biodiversity. But it may seem a stretch to believe that the ecology of the Yorkshire Dales will suffer if the locals start saying 'self' instead of 'sen'.

On the one hand, as was pointed out by James Burridge, some conformity in society is a good thing. On the other hand, distinctive local features, conventions and behaviours can still coexist under a unifying umbrella culture that provides cohesion. Monocultural systems are generally less able to withstand or adapt to changes and stresses than more diverse systems, whether that culture relates to language or local customs or biological diversity. It is like having a toolbox of all kinds of gadgets; the chances of being able to pick the right tool for a job are greater if you have more to pick from in the first place. Arming yourself with just a spanner will ultimately lead to failure when it turns out that you need to tighten a screw instead of a nut.

The fact that linguistic diversity and biodiversity are linked also provides a handy way of telling when either one may be under threat. Biodiversity reflects the complex intermingling of multiple species: animal and insect, vegetable and lichen, fish and fowl and fungi. Individual species can be tracked with greater or lesser degrees of ease, but it is a time-consuming and costly thing to do. The only way a researcher can measure the prevalence of one particular taxonomic group in one particular area is to go there. Now, looking at linguistic markers can actually help ecologists to decide where to focus their attention.

Much of Burridge's data came from the EDA survey, carried out over a smartphone app. Rare species don't use apps, but humans do. Burridge's model could be run on language in any area, and it could even be conducted remotely from the far side of the world without researchers leaving their desks. Smartphones and social media are endless mines of language data that can potentially help us to understand which areas most require environmental protection. The link between biodiversity and linguistic diversity also means that solving a threat to one may solve the threat to the other. Concerted, integrated efforts to maintain both will in addition lead to a better under-standing of how humans interact with ecosystems.

Burridge's work offers real hope for the ways we can care for our habitat, giving us the means to recognise where care needs to be taken, even though at the same time it seems to suggest that some small-scale loss is inev-itable: Isogloss bubbles will always be inclined to merge.

This brings us back to the problem with which I started this chapter – less crucial to the global biosphere but important to me. One day I might be accepted as a York-shireman, not because I've reached the magic 50-year limit, but for the simple reason there is no difference between how Yorkshiremen and Londoners speak. But I hope not; I wouldn't trade my family's distinctive Yorkshire heritage for anything. Even if I still say 'gr-arse' instead of 'grăss'.

Chapter 7

THE PHYSICS OF A MIDLIFE CRISIS

Despite having to spend most of the time sitting in hotel lobbies, I really enjoyed my time at Davos. I found the atmosphere heady and invigorating – both the clean, fresh air that you breathe outside the top-end hotels, and what you find inside. There was a sense of common purpose among the guests and delegates and I came away with a feeling of optimism. Yet under the surface, I was experiencing an unanticipated personal crisis.

I was, at the time, in my early forties. I had been married for almost a decade and had two young sons. I had a good job, nice colleagues, and loving friends and family. I knew objectively that I had so much to be grateful for. Yet something gnawed at me. Being surrounded by such an impressive array of accomplished individuals, seeing so many people who appeared to be more successful, more influential and more self-assured, triggered a corresponding sense of inadequacy. It was an unfortunate application of Newtonian principles: Action had provoked reaction. Despite my years of hard work and dedication, I couldn't help but compare myself to the success of those around me. Their achievements cast a harsh spotlight on my own perceived shortcomings.

While I struggled to articulate this feeling at the time, I realise now what I was experiencing. It was a midlife crisis. As clichéd as it sounds, suddenly, the milestones I had once deemed important – career advancement, financial success, social status – felt hollow. I found myself questioning the choices I had made, wondering if I had veered off course somewhere along the way, and grappling with the unsettling suspicion that I wasn't living up to my true potential.

As I grappled with this minor personal crisis in the weeks and months that followed my time in Davos, the world faced its own. The Covid pandemic erupted, bringing with it widespread political uncertainty and social instability. By May 2023, when Covid-19 was finally declared no longer a global public health emergency, over 760 million people had been infected worldwide, almost 7 million had died, and economies had been stretched to breaking point.

The Covid-19 pandemic confronted us with one of the biggest global risks in living memory. But even at the height of the pandemic, there was hope. In December 2020, less than a year after it began, the United Kingdom was the first nation to begin mass inoculation with a fully tested, clinically authorised vaccine. Other countries soon followed. It was a huge step forward in fighting the virus.

Unfortunately, this also unleashed a new threat on global society – another aspect of the crisis, and one that I found immensely depressing: A vocal minority of the population refused to get vaccinated. Some individuals were worried about potential side effects, including rare cases of fatal blood clots and thrombocytopenia. Some believed mad conspiracy theories. And many could see the benefit of vaccination, but simply thought that if enough people around them got vaccinated, they would be protected by herd immunity and therefore get away with being unvaccinated themselves.

Unfortunately, the more people put their trust in herd immunity rather than taking action themselves, the more misplaced that trust becomes. Between 70 per cent and 90 per cent of the population need to be vaccinated to achieve herd immunity. As the number of unvaccinated persons in a community increases, herd immunity is lost and the risk of vaccine-preventable diseases among the vaccinated rises. There is already evidence that children who refuse vaccination against measles and pertussis – or rather, whose parents refuse it on their behalf – are not only at increased risk from those illnesses themselves, but increase the risk for children who are too young to be vaccinated, or have medical reasons not to be so. The loss of herd immunity puts everyone at additional risk. Decisions based on personal preference have an effect on everyone.

So, anti-vaxxers increase not only their individual risk but also the risk for the whole community. It's what economists call the tragedy of the commons: the tendency of individuals with access to a shared resource to act in their own rather than the common interest, consuming more than their fair share and ultimately leading to everyone being left worse off.

The phrase 'tragedy of the commons' comes from a conundrum that was first articulated in 1833 by English economist William Forster Lloyd. Lloyd was referring to common land where herders could let their cattle graze. Grass does replenish itself, given time, but not if it is grazed too quickly. Simple arithmetic therefore says that if too many cattle graze at once then the commons will be depleted, the cattle will starve, and the herders will lose their livelihood.

If we look at the problem dispassionately, it is clear that to keep grass available for all cattle, the herders should work together to limit the number of grazing animals. What happens, however, is that each herder makes the decision to graze as many of their animals as they can. Individually, the decision is perfectly rational. Collectively, it is disastrous.

The problem was developed in scientific terms by ecologist Garrett Hardin in 1968. Each herder wants to maximise the benefit they gain from grazing their cattle. Whether consciously or unconsciously – and more likely the latter – they pose the question, 'What is the utility to me of adding one more animal to my herd?' That utility has a plus and a minus to it. Call it $+1$ and -1.

On the plus side, the herder has one more well-fed and therefore marketable animal. They will be able to sell it and pocket the proceeds without having to share with anyone else. That is pretty much an entire $+1$.

On the minus side, their animal may contribute to overgrazing. However, the loss of -1 that the herder experiences from this consequence will be shared among all the herders. The herder's personal loss is therefore only a fraction of -1, and it therefore makes sense to add one more animal to their herd. And because the mathematics of utility will work out the same for every individual animal, they will add another, and then another . . .

Unfortunately, every herder with an animal on the commons will be making the same calculation. There is no limit in this estimation to how far they can extend their herd, even though the world in which the herd operates is limited. The only way to prevent it is outside intervention – a legal restriction on what was previously unregulated freedom.

The commons can be something plain and obvious, like grazing land or any kind of natural resource. Or it can be something intangible, like social benefits or the general feeling of goodwill in a community. In the case of Covid-19, that resource was public health. Its depletion represented not only the loss of herd immunity, but the loss of the social good that comes from citizens feeling their health is being protected.

The fact that disease outbreaks occur in clusters is significant, and not just because viruses travel more easily in close quarters. It shows that social influence is an important deciding factor when considering one's actions. The truth is that some people are bad neighbours. Even though vaccinations would ultimately benefit everyone, it's too easy for us to act only in our own short-term self-interest. This is what underpins not only many instances of vaccine refusal, but also littering, overfishing, industrial pollution and many more antisocial practices.

Still, for all those people who act selfishly, there are also people who act selflessly, often out of a sense of – to use a perhaps old-fashioned term – morality. And so the question is: Can physics predict whether such moral behaviour can ever flourish in society? And how can we encourage more people to be less selfish?

As we have seen in previous chapters, trying to understand and predict the behaviour of millions of people in society is incredibly complicated. But there's an old physics joke that provides a new perspective. It's about a dairy farmer who was having trouble getting enough milk from his cows. Nothing he tried worked. Months passed and no one was able to help. In desperation, the farmer mentioned his problem to a theoretical physicist. She agreed to think about the problem and after a week came back declaring she'd found a brilliant solution. The physicist invited the farmer to her laboratory and then drew a large circle on the chalkboard, proclaiming, 'First, we assume a spherical cow . . .'

Physicists aren't known for their jokes but this one is funny (at least for me) because it contains a grain of truth. Physicists take the messy real world and try to make it simple. Of course cows aren't spherical – but if we were trying to calculate, say, the gravitational force between the Earth and a cow, then this assumption would make total sense. When physicists devise physical models, they ignore the variables that have no actual bearing on the problem, and concentrate only on what matters. The gravitational attraction on a spherical cow would be exactly the same as on a cow with the usual body shape and number of heads, legs, tails, horns and udders. However, including all of the latter makes the calculation far more complicated than it needs to be. So, they simplify and assume cows are spherical. In this way, they can test whether their understanding of the underlying physics is correct. In the case of society, we can make similar simplifying assumptions.

Society consists of millions of people who, from a distance, appear to move en masse, almost like a single organism. But in fact, each person is an individual whose choices and actions – whether they recycle, what car they drive, whether they get vaccinated – influence and are influenced by the people they see around them every day.

The same scenario is mirrored in the world of physics. Imagine millions of iron filings on a surface under the influence of a magnet. From a distance, these filings look like a single object, a cloud, shifting and twisting as electromagnetic forces change. But examined more closely, the movement of each filing is most accurately described by looking at the behaviour of its closest neighbours, not by the whole mass. This is called the 'nearest-neighbour' model of magnetism.

Bear with me, and let's assume that every individual in a society is a sphere. Let's lay them out in a giant lattice grid, representing society. Let's now assume each of these spheres has a choice about whether or not it cooperates (gets vaccinated, recycles, doesn't stockpile more toilet rolls than it needs, etc.). Let's assume that non-cooperation leads to a fine, or some other personally undesirable and non-lethal consequence. And let's assume that each sphere only interacts with its nearest neighbours, much like we are all most heavily influenced by those closest to us – our family, friends and neighbours. Under the right conditions, we see that cooperative behaviour can and will take hold in a society.

At first, the inclination to be cooperative or uncooperative is distributed randomly across the grid. Each individual works out how much they stand to gain or lose, in relation to their nearest neighbours, through their behaviour. Based on the success of their neighbours, each individual's strategy is then modified. In effect, individuals come to imitate those who perform better than they do, and waves of imitative behaviour sweep across the lattice.

Translated into the lives of non-spherical humans, people are subject to forces that are akin to physical forces, even if actually they are not. For example, a pedestrian walking down a passage is motivated first of all by the force of their intention to head in a particular direction. Then there may be an attractive force from their intended destination; for example, a coffee bar at break time will attract people from all over. Then they are acted upon, consciously or unconsciously, by a repulsive force from the walls; nothing so exciting as an invisible force field, but simply an awareness that the walls define the passage and that it is best to walk parallel rather than perpendicular to them. And then there are the forces from other pedestrians. Most people find it easier to navigate their way around the members of a crowd rather than barge straight through, and there is an instinct to allow everyone a small measure of personal space, which is effectively a mutually repulsive force. Each of those forces has an effect on the course that the pedestrian will ultimately take; each is a force that can be quantified and measured.

Observations bear out these theories. In a mixed crowd of pedestrians shuffling along the pavement at rush hour, distinct lanes will start to appear in which everyone heads in the same direction. It is even possible for lanes of motion to cross each other with the pedestrians barely breaking step.

This might seem quite intuitive, but the model of breaking everything down into forces has also had unintuitive results. For example, what is the best way to improve the flow of pedestrians in a crowded area? Imagine a square full of people trying to walk in different directions and bumping into each other. Town planners know that the answer to improve flow is to reduce the available space for walking and introduce an obstacle. Less is more. Reducing the space reduces the number of independent options and increases the chances of mutual cooperation, while an obstacle like a pillar in the middle of an intersection increases the likelihood of a rotary flow pattern emerging around it.

So, if a person steps aside to let you pass in a crowded corridor, are they acting morally with conscious volition, or is their gesture purely the result of forces acting on them that they barely recognise? Or is it somewhere in between? Whatever the motivation, the outward effect is of courteous, pro social behaviour.

Looking back at where we started – the Covid pandemic – what actually happened? For the sake of simplicity (spherical cows again), let's divide people into two groups: individualists and collectivists. Individualists generally look after number one; they are the ones who will let their cattle overgraze on the common. Collectivists are more likely to consider how their actions will affect others and adjust them accordingly. Society takes on the attribute of whichever group is dominant.

In their real-world interactions, individualists are also more likely to be solitary by nature. An extreme stereotype is the prepper or survivalist in the backwoods, stockpiling canned food and weapons against the collapse of civilisation. Collectivists tend to be more sociable and to work together more.

Something that became clear early in the pandemic was that the virus spread very easily in crowded conditions.

A sure defence against it was isolation, either of individuals or of small groups of individuals in the form of families. Intuition would therefore say that the virus would spread more in collectivist societies – those where social interactions are closer and more frequent. Individualist societies would automatically be safer.

Quite the opposite turned out to be true. A study of 69 countries – with a total population of 5.87 billion, or some 75 per cent of the world's population – showed that the more individualistic a country's society was, and therefore the more likely its people were to ignore prevention measures like vaccination, the more Covid cases and deaths it had.

It would therefore seem that the best way to promote vaccination is to encourage a collectivist mindset. Every country and culture will have its own best way to go about this, but by considering an individual's interactions with their immediate neighbours, techniques from physics can be used to develop the right combination of incentives to encourage 'selfless behaviour'.

*   *   *

The pandemic was one of the many factors contributing to my own and the world's midlife crisis. There were many more. For the several years previously, a large part of my work had been focused on responding to various crises – from natural disasters like floods and hurricanes, to humanitarian catastrophes like the Syrian refugee crisis or the war in Ukraine. In all these situations the impact on human lives and society can be devastating. I often reflected on how, if there was a way to somehow see the future, to predict where the next conflict was likely to break out or where the next disaster was going to strike, the potential benefits would be enormous. But making such predictions is hard.

There's a human tendency to make future projections in straight lines. Futurists look at present trends and draw straight-line predictions from current data, without considering unexpected twists and turns. We think, 'If this goes on, then . . .' and proceed to make an inevitable prediction that turns out to be far too simplistic because it pays no attention to the continuum of different forces that are all acting and feeding back on one another: political, economic, social, geographical and historical. Conflicts in particular are inherently hard to predict because they arise precisely from a collision between these competing factors.

Still, it's tempting to believe that if only all these forces can be quantified then it should be possible to accommodate all of them into an accurate prediction – and, perhaps, then to tweak them to produce a desired outcome rather than the 'inevitable' one. Starting in the 1940s, science fiction writer Isaac Asimov wrote the *Foundation* series of stories, centred on a fictitious science that he called psychohistory – the ability to predict the rise and fall of whole civilisations. Drawing parallels with the fall of the Roman Empire, the psychohistorians of the Foundation predict the inevitable end of a golden-age Galactic Empire and an anarchic, 30,000-year dark age that will follow. But they also see how, using psychohistory to track and adjust historical trends, the Foundation can narrow that dark age down to a mere millennium, before the Second Empire and an even more golden age arise from the ruins of the first.

It makes for a good story. Sadly, as we have learned throughout this book, the deterministic world that I longed to live in as a boy is not real. The world is a chaotic place where a multitude of forces interact simultaneously, making linear predictions impossible, and where even a small change at one point in time can result in large differences later on. Any inaccuracies in the initial information used in a calculation, no matter how small, can have a dramatic influence on the prediction of future events. This is why even weather forecasts are typically only very accurate for a limited number of days. So, we are still some way off predicting the rise and fall of civilisations.

There does, however, seem to be a way of looking ahead. Rather than trying to stitch together a chain of cause and effect that will disappear into a future haze of chaos ahead of us, the answer is to look backwards, to understand patterns in the past. Patterns tend to repeat, so it follows that by analysing what has happened before, we might be able to predict what will happen again, and avoid the possibility – as mentioned elsewhere in this book – that just asking the question is affecting the outcome. Specifics might still elude us; we may never have psychohistory's ability to pinpoint the rise and fall of whole civilisations, but if we confine ourselves to analysing the past then suddenly things do become clearer.

The patterns might not jump out at us, but they are there, buried in the masses of data that any record-keeping civilisation churns out. And there has never been a richer time in history for gathering that data. More and more newspapers and archives are being digitised, so that searches that would once have taken months in dusty library basements can now be performed quickly and cleanly.

Dr Weisi Guo, then an associate professor in the University of Warwick's School of Engineering, and Sir Alan Wilson, director of special projects at the Alan Turing Institute, took on this challenge to develop their model GUARD (Global Urban Analytics for Resilient Defence). The model draws on the principles of physics and assumes both war and peace to be two stable states. In other words, when they are in equilibrium, a lot of energy is required to change them; when they are out of equilibrium, they will naturally tend to become one or the other very quickly. The researchers then trained an artificial intelligence, which excels at finding patterns in data that are invisible to the human mind, on the latest thinking in complex networks and spatial-interaction theory. Finally they drew on a range of data about phenomena that affect conflict – from geography and culture to political allegiances and military trade partnerships – and fed it to the AI.

The model demonstrates remarkable predictive power. On recent historical data, it achieved between 82 per cent and 94 per cent accuracy in predicting where conflict would break out in peaceful areas and where warring areas will go back to peace, 12 months in advance. It also tells us the combination of conditions that exacerbate the risk of violence.

A much longer-term forecast, without the benefit of AI, was made back in 2010, a decade before my week in Davos. The science journal *Nature* invited a handful of prominent economists, social scientists and historians to make predictions about the decade to come. Peter Turchin, Russian-born but now a professor at the University of Connecticut, accepted the challenge. He wrote a short, 300-word article predicting that 2020 would see levels of social unrest in the United States and around the world not seen since the late 1960s – cities on fire, elected leaders endorsing violence, homicides surging.

The accuracy of his prediction was striking. In 2020, low-level discontent was rumbling all around the globe. In some places it came from government action. In some places it was civil war or insurrection, from people who would very much like to be governments themselves. And sometimes it was discontent from within a populace that simply decided it had had enough. In May, the murder of the African American George Floyd by a white police officer – in front of witnesses, who videoed and replayed it around the world – sparked protests across an already volatile United States and in other countries. The US in particular was a febrile place in 2020. It was gearing up to what would be an unprecedentedly acrimonious presidential election, which would lead to accusations of malfeasance and corruption, and ultimately to a mob descending on Washington DC early in 2021.

There was of course no way of testing his predictions, other than living 10 years and seeing what the world was like when you got there. Having done that, it is hard to argue with the accuracy of his predictions. The events that unfolded after my time in Davos certainly bear them out, both within the United States and on a global scale. Turchin could not have foreseen the pandemic but he would have been the first to say it was not a good year to have one.

Turchin is in fact one of the founders of a field that may be the closest thing we have yet to Asimov's psychohistory.

He calls it cliodynamics – for the Greek muse Clio, goddess of history – and it treats history as a science. Turchin too takes digitised data as his basis. He then looks for 40 social indicators – including factors like economic inequality, population growth, political instability, levels of violence, social mobility and public health metrics – over the range of written human history. After that, he says, it is simply a matter of statistical analysis. The mathematics are not complex, and he only uses ordinary statistical software.

Based on the statistics they find in one set of data, Turchin and his colleagues build a mathematical model.[1] That is when they can be said to 'make a prediction'. They then test that model against other historical data sets to see if the model holds. In effect, they are saying, 'If this model worked *then*, did it work *then*? And what about *then*?' From that, they can build up a reasonable prediction of what *will* happen.

What inspired Turchin's *Nature* article was that he had found a pattern of social instability that has shown up in any civilisation for which records are available. These include the Roman Empire, dynastic China, medieval Europe . . . and the United States, a federal, democratic

---

[1] 'Peter Turchin: The Magnetism of Mathematical History', Enza Jonas-Giugni, *The Science Survey*, 11 January 2021. https://thesciencesurvey.com/spotlight/2021/01/11/peter-turchin-the-magnetism-of-mathematical-history/.

republic a mere 250 years old. Essentially, he detected century-long waves of instability, which he calls Secular Cycles, and superimposed on each wave is a 50-year cycle of violence. Turchin's model looks at any year and tells you roughly where in the latest cycle it is.

Secular Cycles follow a regular pattern. First, a population grows beyond its capacity to be productive. Thus there is a disproportionately large number of young people, with falling wages for those who are meant to be looking after them, and an increased state spending deficit. But that is the background to instability, not necessarily the cause. Populations can stay poor for a long time without it leading to violence. The tipping factor is what Turchin calls 'elite overproduction'. The elite class grows along with all the others, either the old-fashioned way through breeding, or because new opportunities create new ways in. For example, a university education can become available to a new generation that would previously have gone straight into manual work, or economic opportunities can lead to a newly wealthy middle class. However, the number of elite positions – political appointments, for example – stays the same. Hence a growing class competes for a limited number of positions, and that is what destabilises the state.

The historical pattern begins in agrarian societies but carries on into the industrial age, which suggests that the kind of society is immaterial. For example, excess population growth is no longer likely to lead to mass starvation, as it once would have in an agrarian society; instead, it means too many people for too few jobs, leading to unemployment, leading to instability by another route.

As for the 50-year cycles of violence, these are the release of pent-up pressure, acting like a social regulator valve. Social inequality builds up over the decades to the point where something has to give. The release may be through reform, or revolution, or both. For a while thereafter, there is peace. Then either those changes are reversed over time or a society finds new ways of being unequal, and the process begins again. There is no cast-iron rule as to why a Secular Cycle should last 100 years, or the cycle of violence last 50. Those are simply the patterns, give or take, that emerged from the data.

The whole thing – bringing it back to physics, and again, like a regulator valve – is a matter of feedback between systems. To illustrate, Turchin points to another pattern – the boom-and-bust cycles of prey and predator populations in nature. A boom in the mouse population will mean more food for the weasels, so they too will have a boom. There will then be so many weasels that they eat all the mice, whereupon the weasel population starves, leading to a mouse boom as they react to the predator-free environment . . . and so on, round and round. Weasels would do well to study the tragedy of the commons.

Context is also important. Inequality in 18th-century France led to the French Revolution, while inequality in Great Britain around the same time led ultimately to some far-reaching reforms. This was partly because the situations of the two countries were not identical, partly because one country had the example of the other to serve as a warning, and partly because of differences in the methods of government and how it responded to crises. There might always be a time of unrest, but more enlightened policies and a swifter response to the underlying discontent can reduce the pressure and prevent revolution. As we said before, predictions can't be too specific. Turchin is predicting the kind of thing rather than the precise event.

The sad fact is that the kind of thing tends towards a violent outcome.

How did this enable Turchin to make such specific predictions about the next 10 years for the United States? Essentially, he detected the perfect trifecta that has led to unrest in all the other situations he had surveyed: a ruling class growing faster than the number of positions for their members to fill; declining living standards among the general population; and a government that couldn't cover its financial positions. Wages in the United States were being driven down as the number of available workers exceeded the number of available jobs; average wages had stagnated ever since the 1970s, even while

GDP climbed. The elite population was growing, and becoming wealthier: Per head of population, the numbers of doctors and lawyers qualifying between the 1970s and 2010 nearly trebled. And government debt was increasing. The net result was that tensions and inequality soared. It was a process that Turchin could have mapped onto other societies as far back as records appear.

By treating history as a science and spotting the patterns of the past, Peter Turchin was able to predict the future. Why does this matter? His cliodynamics model offers government leaders a playbook for minimising the future risk of civil unrest by understanding the changes that typically lead to instability. While demographic shifts, such as elite overproduction and declining living standards, play a significant role in driving unrest, government actions can exacerbate or mitigate these underlying pressures. Therefore, by recognising these patterns early, governments have the opportunity to implement policies that address the root causes of instability, potentially breaking the cycle of violence and preventing future crises.

* * *

We have come a long way from talking about midlife crises. But perhaps the one thread running through all the scenarios discussed in this chapter, including my own midlife wobble, is our unconscious reaction to the forces that shape our society. Some of these forces may be invisible and difficult to quantify, such as social norms, cultural values and psychological influences. However, whether you're a collectivist or an individualist, their effect can be measured in an analogous way to physical forces like gravity or magnetism. For example, the influence of peer pressure, the impact of media narratives, or the sway of charismatic leaders are all powerful but often intangible forces that shape our behaviours and societal trends.

Turchin's theory suggests that societies, much like individuals, go through life cycles. Just as a person may experience a midlife crisis, societies too experience periods of intense upheaval and transformation, which can be seen as a kind of midlife crisis in their Secular Cycles. Recognising these patterns can help us understand and navigate the turbulent phases both in our personal lives and in the life cycles of societies.

In the months that followed my time at Davos, many of the physics lessons from the previous chapters ran through my mind. As we know, the principle of uncertainty suggests that there are inherent limits to our ability to simultaneously measure the position and velocity of a particle with precision. Perhaps, during a midlife crisis, embracing uncertainty can be liberating. You don't know where you are going, and you don't even know where you are a lot of the time, but it's natural for identities to evolve and it's okay not to have all the answers.

Another concept we have learned is the idea of entropy. During a personal crisis, one feels a similar sense of disorder. However, just as entropy ultimately leads to equilibrium in a closed system, perhaps embracing the disorder can be a precursor to personal growth and settling into a new life phase. The concept of relativity says that the perception of time and space is relative to the observer's frame of reference. So, looked at another way, what may seem like a disaster from one vantage point may be an opportunity for growth.

Reflecting on the turbulence of the years following my time at Davos, these principles feel not just theoretical but deeply practical. The pandemic and the accompanying social upheavals were a bit like a collective midlife crisis for our society, forcing us to confront uncertainties, adapt to new norms and re-evaluate our priorities. For me, it was this period of questioning that arguably inspired me to write this book. I hope that, just as entropy leads to a new equilibrium, perhaps the disorder of the past few years has set the stage for a more balanced and reflective period to come.

Chapter 8

LIFE, THE UNIVERSE AND EVERYTHING

It's been over three decades since my parents bought me that copy of *The Hitchhiker's Guide to the Galaxy*. It set me off on the path of physics, a subject that, at the time, I thought represented the salve to the confusion that I was feeling – a world that was ordered, a world that just made sense. That was my assumption at least. But 30 years later, if I've learned one thing it's this: I was wrong.

The ideas shared in this book show that the world we live in is far from deterministic. We have learned how predicting the movement of just three bodies in space is a near-impossible task because even the tiniest changes in their initial starting position can have a dramatic impact on where they end up. We have learned the baffling, but nonetheless true, fact that subatomic objects can be both waves and particles, and that hypothetical cats in boxes can be both dead and alive at the same time. We have learned how, even in the vacuum of space, tiny fluctuations can appear seemingly out of nowhere, leading, billions of years later, to the formation of stars and galaxies and ultimately life. We have learned how Newtonian mechanics is incapable of describing the behaviour of phenomena like heat or energy or fluids, requiring an approach based, not on precision, but rather probabilities. And we have learned how systems can transition from stable to unstable in the blink of an eye.

I used to think that people were messy and that physics was methodical. But the more I learn about the physical universe, the more I realise that this is a false distinction. The physical universe, while beautiful and wonderful, is just as messy, chaotic and uncertain as the real world of people and society. And it's this fact that makes the achievements of physicists over the centuries all the more remarkable. To make sense of the physical world, physicists had to develop entirely new ways of thinking – from quantum mechanics, to cosmology, to electromagnetism, to thermodynamics, to complexity theory and more – capable of explaining the weird and remarkable phenomena they observed.

And herein lies perhaps the single most important idea in this book. Over the last 300 years, the pioneers of physics developed models that were so rich, so powerful, so nuanced that they are now able to explain not just the physical world but aspects of the social world too. So, for example, while it may not have been their intention, the inventors of quantum mechanics developed models capable of explaining not just the uncertain behaviour of electrons and protons, but also the behaviour of any number of other situations where uncertainty is a factor – including irrational human decision-making. And perhaps that's why over the last 30 years, the lessons and ideas from physics have helped me in so many different ways. And why the same lessons that helped me personally are now being applied in new and surprising ways to benefit society at large, helping us better understand and tackle problems in the real world.

Of course, physics does not have all the answers. This book contains elegant ideas that attempt to make sense of *almost* everything. The themes covered – from our insatiable greed for fossil fuels to the seemingly unstoppable growth in wealth inequality to volatile election outcomes to social unrest – each represent serious threats to our society, our politics and humanity. They do not, however, represent an exhaustive list of all the world's problems. The World Economic Forum's Global Risks Report, which I mentioned in the introduction, catalogues dozens of different risks every year. While this book certainly covered a number of the biggest, it did not cover them all.

Equally, while I have tried to provide a broad-ranging view of physics topics, the scope is not exhaustive, and nor does it cover physics of the future. New advances are being made every year. There are areas of forward-thinking research that do not as yet have firm answers, but might still play a major part in our understanding of the world over the next 30 years.

The concept of the multiverse, for example, is a familiar one in science fiction. Contrary to how it is sometimes portrayed, the multiverse is not the idea of an infinite range of possibilities, like a world in which the Allies lost the Second World War. Rather it is the notion of a finite range of universes that contain not only everything that is, but all possible states of being that don't immediately cancel themselves out for being self-contradictory. The idea is untestable with our present state of knowledge. But, if validated, the multiverse concept could reshape our understanding of societal dynamics. Imagine the implications of interconnected universes on decision-making. The idea that every possible outcome of a decision – whether, for instance, we tackle global warming or inequality or social unrest – plays out in a separate universe might encourage a more nuanced and open approach to societal choices. It could inspire societies to explore diverse options, recognising that different paths lead to various outcomes in the multiverse. Perhaps it would even change how people think about right and wrong, and what constitutes 'the good life'.

Or take the Big Bang. Following that singular event 13.8 billion years ago, we know that the universe is still expanding. Except that the rate of this expansion is slower than it should be. Given the amount of matter we are able to observe, there is not enough gravity to explain the observed rate of expansion. The hypothesised answer to this is dark matter: matter that has gravitational effects but does not interact with light. Dark matter is not just ordinary matter that you can't see. It is a different kind of matter altogether and could constitute most of the universe's mass. If somehow it could be harnessed then its properties could revolutionise our understanding of energy – and indeed of much of the physics in this book.

And why is the universe expanding in the first place, when the natural effect of all that gravity – from both dark and conventional matter – should be to pull it back together again? It is more than just momentum from the initial violence of the Big Bang pushing things apart. Physicists hypothesise that the universe's expansion could be driven by something called dark energy – a repulsive force that drives the universe's accelerating expansion, like similar magnetic poles pushing each other apart. Several times in this book I have shown how thinking about physical forces can help understand invisible, societal forces. Dark energy could be another of those – a force to apply metaphorically to social dynamics, recognising how things are pushed apart as well as how they like to cohere. Understanding the dark energy of societal progress might inspire strategies for fostering collaboration, innovation and global advancement.

We are able to understand the behaviour of the cosmos using Einstein's theories of special and general relativity. These theories work beautifully at large scales but break down when things get really small. At the same time, quantum physics is brilliant at explaining things at the subatomic level but at the large scale it stops working. Physicists, naturally, cannot let this lie and the field of quantum gravity is one result: an attempt to explain large-scale, macroscopic gravitational physics in terms of small-scale, microscopic quantum phenomena. Tying the two together could help bring the power of quantum physics to the fore of large-scale applications, potentially revolutionising information processing and communication through quantum computing. We've looked at how traditional physics can help understand complex social systems, helping us develop better and more effective models for governance and policy-making. But a greater understanding of the very fabric of space-time at a quantum level could help us go much deeper, right to the very core of how the world works and the way that the humans in it interact.

Quantum physics may even underlie the nature of consciousness itself. This was a theory developed by physicist Roger Penrose and anaesthesiologist Stuart Hameroff. We know that our brains contain neurons and that their combined activity is believed to generate consciousness. We also know that neurons contain microtubules that transport substances to different parts of the cell. Penrose and Hameroff theorised that these microtubules are structured in a fractal pattern. Fractals are those beautiful nested patterns in nature that repeat themselves infinitely, from the florets of a cauliflower to the bronchi of our lungs. They are neither two- nor three-dimensional, but are instead some fractional value in between (a statement entirely typical of quantum mechanics), and they achieve the feat of having a finite area inside an infinite perimeter.

A fractal structure to those microtubules, Penrose and Hameroff argued, would enable quantum processes to occur as tiny particles move along them within the brain's neurons. While Penrose and Hameroff's theory was originally met with scepticism, new thinking is reviving their ideas. Cristiane de Morais Smith, a Brazilian theoretical physicist, and Chinese quantum physicist Xian-Min Jin carefully arranged electrons into a fractal pattern and injected photons – particles of light, when light isn't acting as a wave – into this shape to see how they moved across it. They found that the spread of light across the quantum fractal was, sure enough, governed by quantum rather than classical laws. This therefore could describe how particles travel in the neurons around our brains. If thought is based on quantum principles, this could lead to revolutionary advancements in neuroscience revealing the very nature of what it means to be conscious.

All of the theories are speculative – but they represent intriguing possibilities for the future. As physics continues to reveal the hidden mysteries of the universe, my hope is that it may also provide new ways of thinking about the world's problems. A wider range of thinking implies a wider range of solutions.

Pierre-Simon Laplace believed that if you could know everything about the universe at any moment in time then you could in theory wind the clock forward and have perfect knowledge of the future. 'For such an intellect, nothing would be uncertain and the future just like the past would be present before its eyes.' Sadly, or perhaps fortunately, the world around us is not that straightforward Laplace's vision may therefore remain an impossible dream. But the fact that physicists are still trying is a tribute to why I still love the subject.

Ultimately, the ideas in this book have acted as a guide all the way from being a socially awkward schoolboy, to a callow graduate, to where I am now. I never quite found the 'answer to life, the universe and everything' that Douglas Adams had promised, but along the way, physics has above all taught me to embrace the questions.

As another great science fiction writer, Terry Pratchett, once put it, 'every discovery increases the boundaries of ignorance'. And I have learned to savour what he calls 'the strange warm glow of being much more ignorant than ordinary people'. I hope you will join me in it.

ACKNOWLEDGEMENTS

This book would not have been possible without the teachers and professors who inspired and nurtured my passion for physics: Peter Sammut, David Andrews and Todd Huffman. I am also indebted to the physicists and scientists featured in the book who were so generous with their time: Robert Ayres, Adrian Bejan, Peter Bruza, James Burridge, Ted Lewis and Alex Siegenfeld. I am very grateful to my friends, family and colleagues who read various drafts of the book, providing invaluable feedback. They are: Helen Bach, Stephen Battersby, Patrick Baxter, Ayesha Bharmal, Ben Carter, Ginny Carter, Jackie Carter, Yonca Dervişoğlu, Mark Edwards, Charles Emmerson, Hamish Fraser, Tom Hall, Lee Hunter, Ben Jeapes, Erling Kagge, Stuart McLaughlin, Patrick Thomas, Roddy Vann, Ed Wallace, Anna Wishart and Trevor Wishart. Finally, thank you to my incredible agent Peter Tallack, talented editor Sarah Braybrooke at Ithaka, and publisher Jen Gauthier at Greystone for their guidance, support and endless patience.

ABOUT THE AUTHOR

Zahaan Bharmal read Physics at the University of Oxford. His early career was spent as a policy adviser and speech-writer for the British Government and the World Bank. He studied as a Fulbright Scholar at Stanford University where he earned an MBA. Since graduating, he has worked for Google, based in London and Silicon Valley, and is currently a senior director of strategy.

Outside work, Zahaan writes about science for *The Guardian* and has won NASA's Exceptional Public Achievement Medal for services to science communication. He is a trustee of the National Autistic Society and lives in Yorkshire with his wife and two young sons.